INFORMATION HANDLING IN ASTRONOMY – HISTORICAL VISTAS

ASTROPHYSICS AND SPACE SCIENCE LIBRARY

VOLUME 285

INFORMATION HANDLING IN ASTRONOMY – HISTORICAL VISTAS

Edited by

ANDRÉ HECK
Strasbourg Astronomical Observatory, France

KLUWER ACADEMIC PUBLISHERS
DORDRECHT / BOSTON / LONDON

A C.I.P. Catalogue record for this book is available from the Library of Congress.

ISBN 978-90-481-6245-1 e-ISBN 978-0-306-48080-5

Published by Kluwer Academic Publishers,
P.O. Box 17, 3300 AA Dordrecht, The Netherlands.

Sold and distributed in North, Central and South America
by Kluwer Academic Publishers,
101 Philip Drive, Norwell, MA 02061, U.S.A.

In all other countries, sold and distributed
by Kluwer Academic Publishers,
P.O. Box 322, 3300 AH Dordrecht, The Netherlands.

Cover illustration: *"The Emergence of Knowledge" (2002) by Corinne Gerling*

Printed on acid-free paper

Printed in the Netherlands.

This book is dedicated
to the memory
of Gisèle Mersch (1944-2002)

Table of Contents

FOREWORD

This book is dedicated to the memory of Gisèle Mersch whose life ended prematurely in June 2002. Back in the 1970s, when few people were using them, Gisèle introduced me to the arcane secrets of then advanced multivariate statistical methodologies.

I was already involved in more classical statistical studies undertaken at Paris Observatory with Jean Jung: developing and applying maximum-likelihood algorithms to stellar photometric and kinematic data in order to derive absolute luminosities, distances and velocities in the solar neighborhood.

But what could be envisaged with those methodologies was something of another dimension: for the first time, I could really see how to extract information from massive amounts of data without calling for elaborated physical or mechanical theories.

Several pioneering applications were developed under Gisèle's guidance and with her collaboration to study the delicate interface between spectroscopic and photometric data. Thus errors in spectral classifications were investigated as well as predictions of spectral classifications from photometric indices (see Heck 1976, Heck *et al.* 1977, Heck & Mersch 1980 and Mersch & Heck 1980), with very interesting results for the time. Gisèle also took part in studies of period determination algorithms (see Mersch & Heck 1981, Manfroid *et al.* 1983 and Heck *et al.* 1985).

Gisèle's generosity, patience and dedication were impressive. She had set up a statistical consultancy service for the other departments at the University of Liège, Belgium. She would often tell the following anecdote which is full of lessons worthy of considerations by students.

One day, she was approached by someone from the human sciences. That gentleman, who obviously knew little of the elementary mathematical problematics, brought her a case study with n observations and m unknown 'parameters' to be determined, with $n < m$. Gisèle kindly explained him that, in such a situation – less observations than variables – she could not do anything. He had to collect a bigger sample of observations if he wanted the case to be solved. How could she dare! He started threatening to file a complaint with her boss and even higher up in the University if she was to persist in such a non-cooperative attitude.

Shared between offence, compassion, and a strong need to laugh, Gisèle kept however her best face and said that, in such conditions, she had indeed no choice. She invited the arrogant gentleman to come back a couple of days later. After he left, it took her five minutes to write a short Fortran program printing in huge characters on one of those large pages of the computer printout in usage at that time:

"The case has too many unkown parameters
for the number of observations.
It cannot be solved."

You have certainly guessed the end of the story. When he came back, the gentleman had no difficulty to accept the verdict of the machine. It was pure truth since the computer had said it.

Also for students, we often took as an example the paper by Heck *et al.* (1977) where four mathematicians working in different disciplines (astrophysics, medicine, psychology and statistics) collaborated efficiently on a single project: once agreement on the vocabulary used had been reached (for instance, the term 'parameter' did not mean initially the same thing for everybody), the intellectual processes and statistical procedures were the same whether the individuals dealt with were stars, cancer patients or laboratory rats.

Those investigations were expanded later on and other methodologies were investigated with other partners (see *e.g.* Murtagh & Heck 1987, Heck & Murtagh 1989 and Heck & Murtagh 1993), always with the same fascination Gisèle had lit up. Such studies were the forerunners of today's data mining and knowledge building methodologies.

It should be kept in mind that these were never intended to replace physical analysis. They should be seen as complementary, useful to run rough preliminary investigations, to sort out ideas, to put a new ('objective' or 'independent') light on a problem, or to point out aspects which would not come out in a classical approach. Physical analysis is necessary to subsequently refine and interpret the results, and to take care of the details. Nowadays, with many 'virtual observatory' projects dealing with huge amounts of data, those intellectual investments of the past are more than ever justified.

*****

This book completes, with emphasis on history, an earlier volume entitled *Information Handling in Astronomy* and published in the same series (Heck 2000).

After a few considerations by the Editor on the evolution of astronomical data and information handling methodologies in the second half of the last century, E. Biémont reviews how the measurements of time, a fundamental parameter for our science, evolved over ... time.

Several chapters are then devoted to astronomical data processing, starting with a personal account by R. Albrecht followed by contributions centered on specific systems: IHAP (P. Grosbøl & P. Biereichel), FITS (E. Greisen), MIDAS (K. Banse) and AIPS (E. Greisen).

We then move to publications-oriented chapters, by H.A. Abt (Editor) and B. Corbin (Librarian) while G. Eichhorn recalls the development of the Astronomy Digital Library. Next, A. Heck reviews the evolution from early 20th century directories to current online yellow-page services.

Two chapters then deal with education, first by J.E. Bishop on pre-college astronomy education in the US, then by C.C. Petersen on the rôle of planetariums.

Then A. Batten and D. McNally discuss the changing rôle of the *International Astronomical Union* in providing and organizing information, followed by D.H.P. Jones discussing a sometimes controversial matter: the impact of the *Carte du Ciel* project on the development of astrophysics in Europe, and thus on the collection of related data on that continent. The book concludes with a review by S. Dunlop of amateur data and discoveries in the 20th century.

It has been a privilege and a great honor to be given the opportunity of compiling this book and interacting with the various contributors. The quality of the authors, the scope of experiences they cover, the messages they convey make of this book the natural complement of the first volume.

The reader will certainly enjoy as much as I did going through such a variety of well-inspired chapters from so many different horizons, be it also because the contributors have done their best to write in a way understandable to readers not necessarily hyperspecialized in astronomy while providing specific detailed information, as well as plenty of pointers and bibliographical elements. Especially enlightening are those 'lessons learned' sections where authors make a critical review of the experience gained.

It is also a very pleasant duty to pay tribute here to the various people at *Kluwer Academic Publishers* who quickly understood the interest of such a volume and enthusiastically agreed to produce it.

Special thanks are due to Artist C. Gerling whose 'Emergence of Knowledge' (2002) illustrates the cover of this volume.

André Heck
Picos de Europa
November 2002

References

1. Heck, A. 1976, An Application of Multivariate Analysis to a Photometric Catalogue, *Astron. Astrophys.* **47**, 129-135.
2. Heck, A. (Ed.) 2000, Information Handling in Astronomy, Kluwer Acad. Publ., Dordrecht, x + 242 pp. (ISBN 0-7923-6494-5).
3. Heck, A., Albert, A., Defays, D. & Mersch, G. 1977, Detection of Errors in Spectral Classification by Cluster Analysis, *Astron. Astrophys.* **61**, 563-566.
4. Heck, A., Manfroid, J. & Mersch, G. 1985, On Period Determination Methods, *Astron. Astrophys. Suppl.* **59**, 63-72.
5. Heck, A. & Mersch, G. 1980, Prediction of Spectral Classification from Photometric Observations – Application to the uvbyβ Photometry and the MK Spectral Classification. I. Prediction Assuming a Luminosity Class, *Astron. Astrophys.* **83**, 287-296.
6. Heck, A. & Murtagh, F. 1989, Knowledge-Based Systems in Astronomy, Springer-Verlag, Heidelberg, ii + 280 pp. (ISBN 3-540-51044-3)
7. Heck, A. & Murtagh, F. 1993, Intelligent Information Retrieval: The Case of Astronomy and Related Space Sciences, Kluwer Acad. Publ., Dordrecht, iv + 214 pp. (ISBN 0-7923-2295-9)
8. Manfroid, J., Heck, A. & Mersch, G. 1983, Comparative Study of Period Determination Methods, in *Statistical Methods in Astronomy*, **ESA SP-201**, 117-121.
9. Mersch, G. & Heck, A. 1980, Prediction of Spectral Classification from Photometric Observations – Application to the uvbyβ Photometry and the MK Spectral Classification. II. General Case, *Astron. Astrophys.* **85**, 93-100.
10. Mersch, G. & Heck, A. 1981, Preliminary Results of a Statistical Study of Some Period Determination Methods, in *Upper Main Sequence CP Stars*, 23rd Liège Astrophys. Coll., 299-305.
11. Murtagh, F. & Heck, A. 1987, Multivariate Data Analysis with Astronomical Applications, Kluwer Acad. Publ., Dordrecht, xvi + 210 pp. (ISBN 90-277-2425-3).

HALF A CENTURY OF INTENSE MATURATION

A. HECK
Observatoire Astronomique
11, rue de l'Université
F-67000 Strasbourg, France
heck@astro.u-strasbg.fr

Abstract. The 20th century, and especially its second half, has seen a dramatic change in the way data were collected, recorded and handled, as well as how the ultimate product was distributed either to scientists, to students or to the public at large. Beyond a compact historical review, this paper offers also a few considerations touching issues such as the available manpower and the place of astronomy in our society.

1. From Freezing in the Domes ...

That mountain gear had been bought in the early seventies at a well-known sports shop downtown in Paris' *Quartier Latin.* It was a *must* for a young astronomer who was going to visit observatories round the world. Much of astronomical observing was still then carried out from within the domes, with an inside temperature equal to the outside one in order to avoid air turbulence through the opening (that would blur images). In deep winter, this meant freezing for twelve hour periods.

So in order to survive, it was necessary to look like a Michelin *Bibendum* dressed in that mountain gear complete with lined shoes, thick trousers and hooded jacket stuffed with bird down. Only the fur gloves would be temporarily taken off for the necessary operations with the hands and then put on again.

That equipment was so cosy and warm that it must have happened at least once to every astronomer and night assistant of the time to fall sound asleep in the loneliness and darkness of the dome, occasionally with the help of a gentle music. Under the sky lurking through the dome opening, the telescope drive was then left to itself, gently steering the instrument

A. Heck (ed.), Information Handling in Astronomy – Historical Vistas, 1-13.

out of that opening[1]. Also, without precise guiding, the objects pointed at would then be drifting out of the spectrograph slits or the photometer diaphragms, or leaving potatoid and trailed images on Schmidt plates ...

Yes, this happened even to the best ones (but do not expect them to brag about it) and generally during the weakest part of the night or while digesting the midnight meal. Nights were long in winter, observing runs were sometimes very long too (occasionally lasting one full month, something unimaginable today), and sleeping hours during the day were not many: it was necessary to review daily all the work done during the previous long night and to prepare the next long one.

If still in the seventies that beloved mountain gear was a bulky, albeit not so heavy part of the luggage when travelling to observatories round the world (Fig. 1), it was not going to be so for very long.

Thanks to the development of detectors, computers, electronics and communications, astronomers would be progressively and almost totally removed from the domes, spending their observing sessions in air-conditioned rooms, not only with light and comfortable seating, but also with facilities at hand for real-time or quick-look analysis of the collected data. Rapidly, all these became digitized and recorded on magnetic media. At the same time, things would also be influenced from up there, high above ground, by space-borne instruments.

2. ... to Novel Observing and Data Handling

The *International Ultraviolet Explorer (IUE)*[2] (see Fig. 2), launched on 26 January 1978, has been the first space-borne instrument welcoming visiting astronomers in real time, just like most ground-based observatories, with the difference that the telescope was not in an adjacent dome, but in a geosynchronous orbit over the Atlantic Ocean. It was shut down on 30 September 1996 after 18 successful years of operations (while its expected lifetime was three years), having become by then the longest astronomy space mission with more than 100,000 observations of celestial objects of all kinds, ten dedicated international symposia and more than 3,500 scientific papers at the time it was turned off. A fantastic achievement for a 45cm telescope.

In many respects, IUE has been the precursor of modern astronomical observing. Integral to the satellite exploitation were the strict procedures, such as those for spacecraft handover between the two ground stations op-

[1]Very few were then the domes equipped with servo-mechanisms coupling telescope and dome slit movements.

[2]For details on the *International Ultraviolet Explorer (IUE)*, see for instance the eight post-commissioning papers published in *Nature* **275** (5 October 1978) and the commemoration volume edited by Kondo *et al.* (1987).

Figure 1. Astronomers J. Manfroid (left) and A. Heck moving down from Mount Chiran station of Haute Provence Observatory on 21 December 1979: dressed in mountain gear of the time (also used for observing), radio in backpack, luggage on sledge topped by snowshoes needed over deep snowdrifts. Photometric data were still collected in analog mode there. Located at an altitude of about 1900m on the first ridge of the Alps from the West, built in 1974 by French CNRS and decommissioned from professional observing in 1986, that station is now in the hands of a local educational association.

erating it (GSFC in the US and Vilspa in Europe), as well as the chains of commands and responsibilities needed in space operations for the instrument safety and for the efficiency of observing: visiting astronomer, resident astronomer, telescope operator, spacecraft controllers monitoring also communications and computer center, plus overall permanent IUE control at NASA.

People realized that those procedures used for a spacecraft in geosynchronous orbit at some 36,000km from the Earth could be applied for remotely piloting a telescope at "only" a few thousand kilometers distance somewhere on Earth – saving travel money, substantial travel time, time difference disturbance and fatigue to the observers.

They also realized that the assistance provided to visiting astronomers through the team of resident ones, as well as the flexibility and dynamics introduced in the scheduling, for targets of opportunity and service observing for instance, could be extrapolated to ground-based instruments for optimizing their return (see *e.g.* Robson 2001). Additionally, with the panchromatization of astronomy and the multiplication of joint observing campaigns (see *e.g.* Peterson *et al.* 2001), procedures were progressively generalized and standardized for all instruments, ground-based or space-borne.

But more importantly in the context of this book, the space agencies operating IUE (NASA, ESA & SERC) agreed on real data policies which inspired modern astronomical archives avoiding, as has happened too often in the past, data disappearing for ever on the shelves or in the drawers of the original observers – when they were logged at all.

An IUE policy was to declare the data publicly available one year after the corresponding observations had been conducted. This meant too that an *ad hoc* service had to be set up by the agencies, providing access to the data archived. This, in turn, involved sometimes reprocessing large amounts of data, or transfering data to new media as the technology evolved. Living archives were born. Lessons from IUE can also be found in projects for "virtual observatories" (see *e.g.* Benvenuti 2002).

3. A Dramatic and Quick Evolution

It has been an exciting time to be an active part of this evolution, both as a "ground-based" and a "space" observer, but also as a heavy user of big amounts of data for personal research, as a developer of databases, and as an insider in archive/data centers and in their followers.

That evolution from individual records to catalogs, data centers, information hubs and nowadays "virtual observatory" projects has already been dealt with in a chapter of the previous volume (Heck 2000c) where other specific points have been tackled too such as:
– astronomy as essentially as a "virtual" science,
– the structure of the information flow in astronomy (Fig. 3),
– "virtual observatory (VO)" projects,
– success stories (such as CDS'),
– methodological lessons learned,
– the real slot of electronic publishing,
– quality versus automation,
– the need of prospective,
– education and communication,
and so on.
There is no need to repeat here those discussions. Please refer to the paper mentioned as well as to Heck (2002).

A couple of additional comments are however in order considering the historical perspective of the present volume.

Figure 2. Observing with the *International Ultraviolet Explorer (IUE)* at Vilspa on 5 April 1978. From left to right: Telescope Operator F.J. Castro, (then) Resident Astronomer A. Heck, (then) Deputy Observatory Controller M.V. Penston and Visiting Astronomers M. Perinotto and S. Aiello. This was the first European observing run with visiting astronomers after commissioning the spacecraft and its scientific instrument. IUE has been the precursor of modern astronomical observing in many respects (see text), including through the policies applied to the data collected.

4. A Big and Complex "Business" Today

The self-explicit graph on Fig. 3 gives a schematic idea of today's astronomy information flow, from data collection to processed information tuned to various audiences, including internal iterations and input from related disciplines. Such a variety of perspectives is to be found in the present volume and in the previous one (Heck 2000a).

Astronomy has also become a big business as any visitor to the exhibition areas of AAS Meetings[3] (for instance) can appreciate nowadays: big projects for telescopes, arrays, spacecraft, auxiliary instrumentation, not to forget surveys, VOs, and so on.

As pleasantly recalled by Blaauw (2001), 17th-century Johannes Vermeer's *"Astronomer"* did not know all the deadlines we have to meet today, nor the selection committees, nor the referees, nor the financial austerity

[3]AAS = American Astronomical Society (`http://www.aas.org/`).

imposed on university scientific research, and so on. Such a reasonably quiet life was still largerly taking place among our colleagues in the first half of the 20th century.

Many of us have experienced a dramatic evolution over the last decades of the 20th century. Perhaps only the youngest astronomers would not remember how (not so long ago) we were still using mechanical typewriters, speaking to colleagues over noisy phone lines (sometimes hard to connect and frequently breaking down) and how we were dependent, to work and publish, on what nowadays we call the "snail mail". At that time, we happily ignored the e-mail stress, we had no e-boxes flooded daily with hundreds of spams and we were saved from masses of junk mail.

People of my end-of-WWII generation still started working on their thesis with mechanical computing machines and slide rulers. Then came the first computers (see also Albrecht 2003) using tons of punched cards – something today students look at with puzzling anxiety before starring right in your eyes as if they were meeting jurassic remnants in real life.

I still remember the day the first HP pocket calculator was introduced to us at Liège Institute of Astrophysics and when the first IBM 360 became operational at the University Computer Center (monopolizing half the basement of the Institute of Mathematics). The stellar evolution programs of my Liège colleagues, as well as my own maximum-likelihood algorithms, would suddenly take less than entire nights to converge – something done today in a few seconds on my already old portable computer.

At the same time, and because of such increasing computer capabilities, methodologies were developed to deal with bigger amounts of data as well as with textual material. Bibliometry had taken off (see also Abt 2003, Albrecht 2000, Corbin 2003, Eichhorn *et al.* 2003, Lequeux 2000 and Grothkopf 2000).

Education was not left aside. In Liège, at the end of the sixties, L. Houziaux had designed a pioneering machine (Houziaux 1974) to teach astronomy, certainly rudimentary by nowadays standards, but it was a fully working device, complete with sound, slides, multiple choices, steps backwards, etc.

By the beginning of the nineties, the spread of networks and the availability of the World-Wide Web (WWW) had given additional dimensions, not only to work and to communicate, but also to educate and to interact with the society at large (see also Bishop 2003, Madsen & West 2000, Maran *et al.* 2000, Norton *et al.* 2000, Percy 2000, Petersen 2003 and Petersen & Petersen 2000), including active amateur astronomers (*cf.* Sect. 2.5 of Heck 2000d) who benefitted fully of the evolution (see also Dunlop 2003 and Mattei & Waagen 2000).

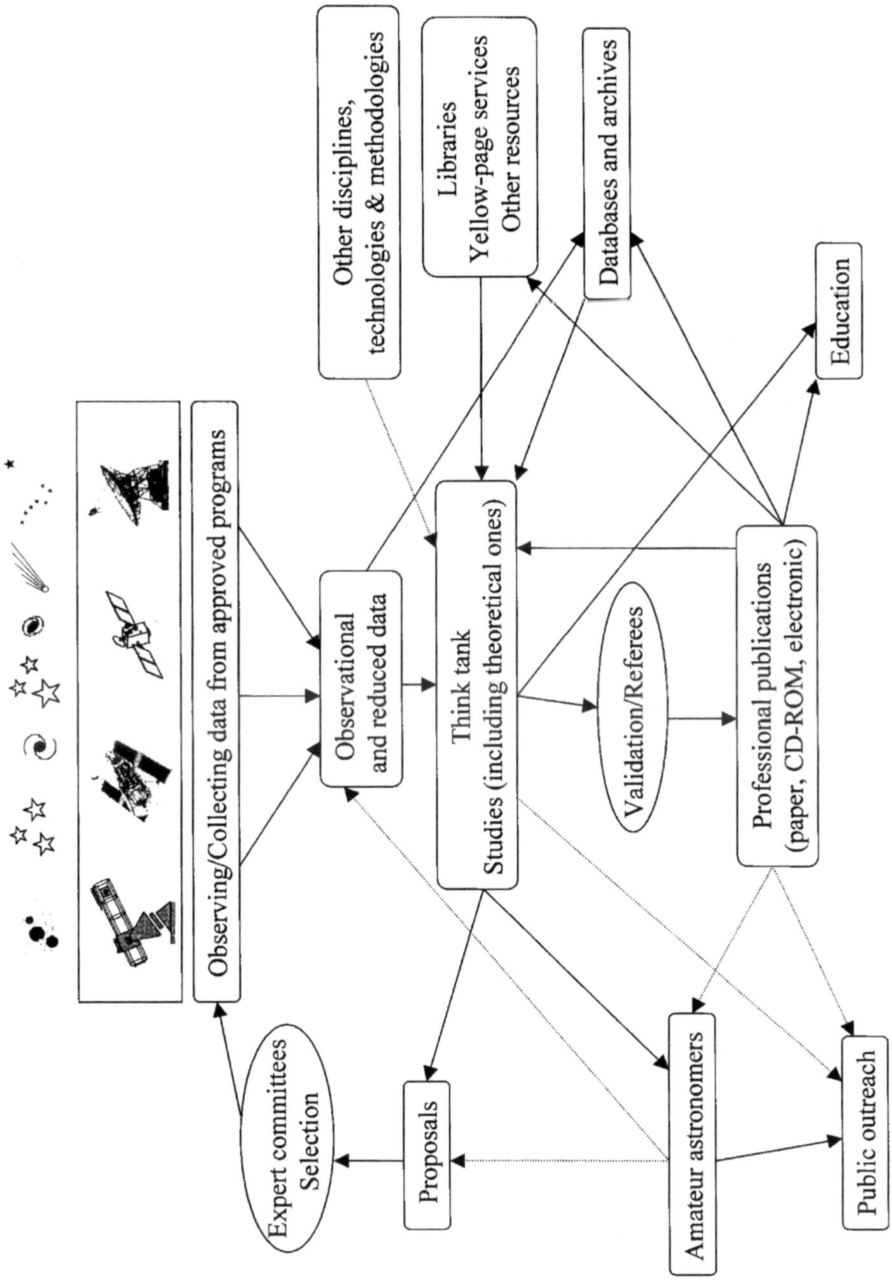

Figure 3. An illustration of the information flow in astronomy (Heck 2000b).

But before the advent of sophisticated information handling methodologies, there was an enormous development and diversification of instrumentation with a surge of momentum in the sixties-seventies which could be illustrated by the series of three conferences co-organized by the European Southern Observatory (ESO) on large telescope design (West 1971), on auxiliary instrumentation for large telescopes (Laustsen & Reiz 1972) and on research programs for large instruments (Reiz 1974).

The media had a parallel evolution. From paper sheets and photographic plates, via punched cards, paper tapes, microfiches[4] and microfilms, to magnetic drums, magnetic tapes of all kinds and disks of all sorts, one simple conclusion is immediate: the medium life is short nowadays!

The 20th century has also been a period when the measurement of time – our sometimes paradoxical reference when diving into the cosmos – evolved dramatically (see *e.g.* Biémont 2003).

Professional associations and, first of all, our world-wide league, the International Astronomical Union (IAU), had also to adapt themselves to the new media and context (see *e.g.* Andersen 2000 and Batten & McNally 2003).

5. "Objectivization" and "Massification" of Information

With its natural intelligence package behind it, the human eye is an exceptional instrument perceiving an extremely large range of contrasts, tones and nuances as any visual planetary observer can testify. People who attended total solar eclipses are also generally disappointed not to find later on, in pictures and movies, the same magnificence they saw when witnessing that fascinating natural phenomenon.

But, as we all know, the human eye has its limitations. First of all – and this is perhaps the most important restriction for us astronomers – it is operating only in the visual range, per definition. Second, its sensitivity is rather limited. We have therefore to assist it by collecting and intensifying devices that, at the same time, are also able to work outside the visual range (radio, infrared, ultraviolet, X-rays, γ rays, ...) and that can be sent outside the turbulent filter of the Earth's atmosphere.

Third, the cerebral firmware behind the human senses has also its complex limitations. It is able to recognize instantly a voice, including its emotional contents – something machines are still largely unable to do efficiently today. But it cannot deal, as fast as computers, with complex calculations or huge amounts of data. Its possible lack of objectivity is another serious issue.

[4]Still remember the microfiches hailed at the beginning of the seventies as The Medium of the Future because of its compactness? How many of us are still using them today?

Therefore data have been progressively recorded through mechanical, analytical, photographical and, of course, always more diversified electronic means. This increasingly removed observational and instrumental biases while improving speed, sensitivity, spectral range, dimensionality and resolution.

Computer and software packages, tools and standards have been adapted to astronomical needs (see *e.g.* Albrecht 2003, Banse 2003, Cheung & Leisawitz 2000, Greisen 2003a&b, Grosbøl & Biereichel 2003, Hanisch 2000, Jacoby & Tody 2000 and Wallace & Warren-Smith 2000), including history-making FITS (Greisen 2003a & Wells 2000). Notes Greisen (2003a): "Our community needs to adopt a more aggressive and inclusive process for standards development".

Earlier concepts, such as the "data flow" one, were given a stricter and more rigorous formulation (Quinn 1996) for an optimum transition of the raw data from the collecting devices to the final product in the hands of the users.

Interoperability of astronomy-related resources has become, more than ever, a critical issue (Genova 2002) with the global integration of those resources in VO projects and others.

Sophisticated algorithms have been progressively developed too in order to deal with bigger and bigger amounts of multidimensional data (including non-quantitative ones) under less and less restrictive conditions. Dedicated conferences have been organized. See *e.g.* Heck (2000c) and Murtagh (2000), as well as the references quoted therein.

We are still a way from W. Gibson's (1986) characters, "jacking in" directly with knowledge bases – if it will ever happen without elaborated assistance compensating the brain complexities mentioned earlier. From the succint and compact historical evolution described above, it should be clear, however, that the profile needed today for a young astronomer is very far from what it was only three decades ago (a trifle, in terms of astronomical timescales), when juggling with slide rulers and expertise with logarithmic tables were among the requirements.

6. In fine

A few final comments might be in order.

6.1. COSMIC TERATOLOGY?

News bulletins rarely speak of trains and planes that arrive on time. Physicians are quite logically interested in illnesses, deviations, abnormalities of all kinds since they have to remove them – as much as possible – from people's lives.

There is no need to run detailed statistics of astronomical research programs and publications to realize that quite a significant part of our activities are devoted to cosmic teratology, *i.e.* to the study of peculiarities, deviations, and so on. Are we however dedicating enough time to the study of "normal" objects? We do not have to cure celestial objects, so there is no real emergency justifying that we neglect thorough investigations of "normalities", needed to build reference sequences, in turn necessary to better understand peculiarities.

Briefly coming back for an example to the IUE satellite, when we were putting together an atlas of ultraviolet spectra of normal stars (Heck *et al.* 1984), most selection committee members recognized the importance of the program (and most used the atlas subsequently), but the pressure was so strong for observing non-normal objects that it has been really difficult to obtain the observing shifts needed for completing the samples of normal spectral sequences. They were systematically given the lowest priorities in terms of time assignment.

Quite naturally, the more we observe objects, the more peculiarities, variabilities, etc., are found – which makes in turn more important the need to define normalities and references. Big projects are not new (see *e.g.* Jones 2003). It would be appropriate upcoming ones dedicate an *ad hoc* fraction of their activities to general cosmic characteristics and properties, and do not concentrate excessively on deviations and peculiarities.

6.2. WHERE MANPOWER MATTERS ALSO

It is said that only 1% of all samples and data from the Moon missions have been analyzed, that about 10% of them have been "looked at", and that the rest has been stowed away, probably for ever.

Have we the same situation in astronomy? Some time ago, I tried to run a survey on the usage of databases and archives in astronomy, but never received exploitable answers. The most plausible reason is that probably database managers do not really have the data to say how much of their holdings have been used (analyzed in details or other) and what percentage led to publications, resp. to advancement of knowledge.

One of the conclusions by Abt (2003) is that: "If we want to increase our output of papers, we should employ more astronomers rather than to build more telescopes".

Although this might not seem related at first sight, I have continually to remind people that the prices of Kluwer's books, including this one, are of the same order as those of any books of the same quality, be they reference works, conributed or edited books, monographs or others[5].

[5]In order to lower their prices (and the inherent risks), other publishers are in prac-

For some mysterious reasons, astronomers always seem to expect to receive things for free or cheap[6]. But exactly because the astronomy community is small, the circulation of professional astronomical publications is small and prices of commercial products cannot be brought down as much as one would hope for.

Increasing manpower in astronomy goes much beyond training more good students. It is directly related to the importance the society is giving to our science today. After the end of the Cold War and long after the landing of man on the Moon, the society at large has now openly other priorities (such as health, environment, security, unemployment, ...) than space investigations or cosmological perceptions.

It is up to all of us, through education, public relations and appropriate representation, to act in such a way our science occupy the rank we believe it should have in mankind's priorities. This is a daily task.

References

1. Abt, H.A. 2003, Changes in Astronomical Publications during the 20th Century, this volume.
2. Albrecht, R. 2000, Computer-Assisted Context Analysis of Databases Containing Scientific Literature, in *Information Handling in Astronomy*, Ed. A. Heck, Kluwer Acad. Publ., Dordrecht, 109-119.
3. Albrecht, R. 2003, Evolution of Data Processing in Optical Astronomy – A Personal Account, this volume.
4. Andersen, J. 2000, Information in Astronomy: The Role of the IAU, in *Information Handling in Astronomy*, Ed. A. Heck, Kluwer Acad. Publ., Dordrecht, 1-12.
5. Banse, K. 2003, The Munich Image Data Analysis System, this volume.
6. Batten, A. & McNally, D. 2003, The Changing Role of the IAU in Providing and Organising Information, this volume.
7. Benvenuti, P. 2002, Some Thoughts about the Virtual Observatory Concept, in *Organizations and Strategies in Astronomy III*, Ed. A. Heck, Kluwer Acad. Publ., Dordrecht, 107-119.
8. Biémont, E. 2003, Evolution of Time Measurement in Astronomy, this volume.
9. Bishop, J.E. 2003, Pre-college Astronomy Education in the United States in the Twentieth Century, this volume.
10. Blaauw, A. 2001, Foreword, in *Organizations and Strategies in Astronomy II*, Ed. A. Heck, Kluwer Acad. Publ., Dordrecht, vii-ix.
11. Cheung, C. & Leisawitz, D. 2000, New Frontiers in NASA Data Management, in *Information Handling in Astronomy*, Ed. A. Heck, Kluwer Acad. Publ., Dordrecht, 45-63.
12. Corbin, B.G. 2003, The Evolution and Role of the Astronomical Library and Librarian, this volume.
13. Dunlop, S. 2003, Amateur Data and Astronomical Discoveries in the 20th Century, this volume.

tice requesting book editors or conference organizers to purchase themselves a minimum number of copies.

[6]This comment could be put in parallel with the discussion by Albrecht (2003) about astronomers abhoring commercial software packages also for some unclear reasons.

14. Eichhorn, G. *et al.* 2003, The Development of the Astronomy Digital Library, this volume.
15. Genova, F. 2002, Interoperability, in *Astronomical Data Analysis Software and Systems XI*, Eds. D.A. Bohlender, D. Durand & Th.H. Handley, *Astron. Soc. Pacific Conf.* **281**, in press.
16. Gibson, W. 1986, Neuromancer, Grafton, London, 318 pp. (ISBN 0-586-06645-4).
17. Greisen, E.W. 2003a, FITS: A Remarkable Achievement in Information Exchange, this volume.
18. Greisen, E.W. 2003b, AIPS, the VLA, and the VLBA, this volume.
19. Grosbøl, P. & Biereichel, P. 2003, IHAP: Image Handling and Processing System, this volume.
20. Grothkopf, U. 2000, Astronomy Libraries 2000: Context, Coordination, Cooperation, in *Information Handling in Astronomy*, Ed. A. Heck, Kluwer Acad. Publ., Dordrecht, 165-174.
21. Hanisch, R.J. 2000, Information Handling for the Hubble Space Telescope, in *Information Handling in Astronomy*, Ed. A. Heck, Kluwer Acad. Publ., Dordrecht, 135-153.
22. Heck, A. (Ed.) 2000a, Information Handling in Astronomy, Kluwer Acad. Publ., Dordrecht, x + 242 pp. (ISBN 0-7923-6494-5).
23. Heck, A. 2000b, Foreword, in *Information Handling in Astronomy*, Ed. A. Heck, Kluwer Acad. Publ., Dordrecht, vii-x.
24. Heck, A. 2000c, From Data Files to Information Hubs: Beyond Technologies and Methodologies, in *Information Handling in Astronomy*, Ed. A. Heck, Kluwer Acad. Publ., Dordrecht, 223-242.
25. Heck, A. 2000d, Characteristics of Astronomy-Related Organizations, *Astrophys. Sp. Sc.* **274**, 733-783.
26. Heck, A. 2002, The Impact of New Media on 20th Century Astronomy, *Astron. Nachr.* **323**, 542-547.
27. Heck, A. *et al.* 1984, IUE Low-Dispersion Spectra Reference Atlas – Part 1. Normal Stars, *European Space Agency* **SP-1052**, 476 pp. + 34 plates.
28. Houziaux, L. 1974, A Teaching Machine for Elementary Astronomy, *Observatory* **94**, 109.
29. Jacoby, G.H. & Tody, D. 2000, The Use of the IRAF System at NOAO, in *Information Handling in Astronomy*, Ed. A. Heck, Kluwer Acad. Publ., Dordrecht, 73-92.
30. Jones, D.H.P. 2003, Was the Carte du Ciel an Obstruction to the Development of Astrophysics in Europe?, this volume.
31. Kondo, Y. *et al.* (Eds.) 1987, Exploring the Universe with the IUE Satellite, D. Reidel Publ. Co., Dordrecht, x + 788 pp. (ISBN 90-277-2380-X).
32. Laustsen, S. & Reiz, A. (Eds.) 1972, ESO/CERN Conference on *Auxiliary Instrumentation for Large Telescopes*, xiv + 526 pp.
33. Lequeux, J. 2000, To be Editor in Chief of a Primary Scientific Journal: From Manual Work to Electronic Publication, in *Information Handling in Astronomy*, Ed. A. Heck, Kluwer Acad. Publ., Dordrecht, 155-164.
34. Madsen, C. & West, R.M. 2000, Public Outreach in Astronomy: The ESO Experience, in *Information Handling in Astronomy*, Ed. A. Heck, Kluwer Acad. Publ., Dordrecht, 25-43.
35. Maran, S.P. *et al.* 2000, Astronomy and the News Media, in *Information Handling in Astronomy*, Ed. A. Heck, Kluwer Acad. Publ., Dordrecht, 13-24.
36. Mattei, J.A. & Waagen, E.O. 2000, Data Handling in the AAVSO: An Example from a Large Organization of Amateur Astronomers, in *Information Handling in Astronomy*, Ed. A. Heck, Kluwer Acad. Publ., Dordrecht, 205-222.
37. Murtagh, F. 2000, Computational Astronomy: Current Directions and Future Perspectives, in *Information Handling in Astronomy*, Ed. A. Heck, Kluwer Acad. Publ., Dordrecht, 121-134.

38. Norton, A.J. *et al.* 2000, Astronomy Teaching at the Open University, in *Information Handling in Astronomy*, Ed. A. Heck, Kluwer Acad. Publ., Dordrecht, 187-193.
39. Percy, J.R. 2000, Astronomy Education: Description, Organization, and Information, in *Information Handling in Astronomy*, Ed. A. Heck, Kluwer Acad. Publ., Dordrecht, 175-185.
40. Petersen, C.C. 2003, The Birth and Evolution of the Planetarium, this volume.
41. Petersen, C.C. & Petersen, M.C. 2000, The Role of the Planetarium, in *Information Handling in Astronomy*, Ed. A. Heck, Kluwer Acad. Publ., Dordrecht, 195-204.
42. Peterson, K. *et al.* 2001, Coordinating Multiple Observatory Campaigns, in *Organizations and Strategies in Astronomy II*, Ed. A. Heck, Kluwer Acad. Publ., Dordrecht, 103-120.
43. Quinn, P. 1996, The ESO Data Management Division, *ESO Messenger* **84**, 30-33.
44. Reiz, A. (Ed.) 1974, ESO/SRC/CERN Conference on *Research Programmes for the New Large Telescopes*, xviii + 398 pp.
45. Robson, I. 2001, New Strategies in Ground-Based Observing, in *Organizations and Strategies in Astronomy II*, Ed. A. Heck, Kluwer Acad. Publ., Dordrecht, 121-137.
46. Wallace, P.T. & Warren-Smith, R.F. 2000 Starlink: Astronomical Computing in the United Kingdom, in *Information Handling in Astronomy*, Ed. A. Heck, Kluwer Acad. Publ., Dordrecht, 93-108.
47. Wells, D.C. 2000, The FITS Experience, in *Information Handling in Astronomy*, Ed. A. Heck, Kluwer Acad. Publ., Dordrecht, 65-72.
48. West, R. (Ed.) 1971, ESO/CERN Conference on *Large Telescope Design*, xiv + 500 pp.

EVOLUTION OF TIME MEASUREMENT IN ASTRONOMY

E. BIÉMONT
IPNAS, Université de Liège †
Sart Tilman
B-4000 Liège, Belgium
and
Astrophysique et Spectroscopie
Université de Mons-Hainaut
Rue de la Halle, 15
B-7000 Mons, Belgium
e.biemont@ulg.ac.be

Abstract. Astronomical phenomena, such as the waxing and waning of the Moon, the succession of days and nights and the pattern of the seasons define a time which is basically cyclical. During many centuries, rather simple devices, such as water clocks or astrolabs and, later on, mechanical clocks, have been used by astronomers for defining realistic but low accuracy time scales. Lately, the atomic time, with its unprecedented precision, has open the way to a more accurate investigation of astronomical phenomena. From cyclical, the time of mankind has become definitely linear and the astronomers seem to have lost its control ...

1. General considerations

We know, since Copernicus, that the Earth revolves around the Sun and rotates along its own axis. The Ancients believed, according to Ptolemy, that the Earth was stationary and that the Sun and Moon moved around it. It is well established now that the Earth is spinning on its axis in an anticlockwise way and is moving in the same direction. In addition, the major celestial bodies of the solar system interact in a very complicated way and an accurate observation of these movements allows to evidence

†The author is Research Director of the Belgian National Fund for Scientific Research (FNRS).

A. Heck (ed.), Information Handling in Astronomy – Historical Vistas, 15-34.

small perturbations of simple motions not recognized in the Antiquity. Elementary and natural units of time (*i.e.* the year, month, week and day) are imposed by the basic astronomical motions of the Earth and of the Moon. Traditionally, the date of an event is composed of two parts: a system of identification of the day, provided by a calendar[1], and a system of subdivision of the day (hours, minutes and seconds). Many lunar, solar or luni-solar calendars have been proposed in the past and most of them try to make compatible – strictly speaking an impossible task! – different natural cycles with well defined astronomical meanings:

- the tropical year which separates two passages of the Sun through the vernal point (spring equinox);
- the lunation or synodic month corresponding to the time interval between two new Moons;
- the real solar day defined by two successive passages of the Sun through the meridian of a given place.

One of the purposes of this chapter is to show how the time measurement has evolved to become progressively more and more sophisticated and accurate. This evolution has been very slow during many centuries but has accelerated considerably during the past few decades. The history of time reckoning is related to the evolution of the techniques and of the ideas but has been frequently characterized by indecisions and sometimes incoherence.

2. Basic astronomical units

2.1. THE TROPICAL AND THE SIDEREAL YEARS

The most obvious way to determine the duration of the tropical year (associated to biological rhythms and changes of vegetation), at Northern or Southern latitudes, is to observe the length of the shadow of a gnomon and to write down the day when it is the shortest at Noon, this day indicating the summer solstice. Observing similarly several consecutive solstices allows to deduce a mean value of the tropical year. Close to the equator, there is no shadow produced at Noon by the gnomon at the solstice, the Sun being located exactly at the zenith. This provides another mean, used in the past by the Incas, to derive a mean value of the tropical year.

Modern astronomers define the mean *tropical year* as the time interval between two successive passages of the Sun through the vernal equinox. This time interval varies from year to year due to nutation and interaction with

[1]The most universal calenderic system is the Gregorian calendar which is essentially solar with the exception of Easter problematics.

other planets and shows a small secular decrease with time[2]. Its present value is 365.24222 d or 365 d 5 h 48 m 48 s.

The astronomers usually define the *sidereal* year as the period of revolution of the Sun in the ecliptic plane with respect to the stars. The modern value is 365.25636 d or 365 d 6 h 9 m 9 s.

2.2. THE SYNODIC AND THE SIDEREAL MONTHS

Months are determined from the motion of the Moon, the average time interval between two successive new Moons defining the lunation. The mean *synodic month*, obtained by observing a large number of cycles, shows a small secular increase with time which is given by the formula:

$$29.5305888531 + 0.00000021621\, T - 3.64\ 10^{-10}\, T^2,$$

T being the date expressed in centuries after AD 2000. A modern estimate is 29.53058885 d or 29 d 12 h 44 m 2.88 s.

It is usual also to define the *sidereal month* as the period of revolution of the Moon in its orbit around the Earth relative to the stars as seen from the Earth: 27.3216609 d or 27 j 7 h 43 m 11.5 s. The *tropical month* corresponds to two successive passages of the Moon through the first point of Aries: 27.3215816 d or 27 j 7 h 43 m 4.7 s.

2.3. THE WEEK AND THE DAY

Since the presemitic time, the number *seven* had a sacred meaning in relation with the number of days of a lunar phase. The division of the month in weeks, naturally associated with these phases, was already valid in Antiquity, namely among the Chaldeans and the Jews.

The day is undoubtedly a natural unit of time for vegetal, animal and human lives. The *mean solar day*, on the one hand, is defined as the average period of rotation of the Earth about its axis with respect to the Sun. Any particular day may vary from the mean value by up to 50 s. The *sidereal day*, on the other hand, defined as the period of rotation of the Earth around its axis with respect to the stars, is shorter by about 4 minutes than the mean solar day: 1 sidereal day = 86164.1 s = 23 h 56 m 4.1 s.

The division of the day in 24 hours and of the hour in 60 minutes is rather artificial and its origin dates back in the Babylonian civilization. All the attempts to entirely decimalize the division of time (including thus

[2]This variation is given by:

$$365.2421896698 - 0.0000615359\, T - 7.29\ 10^{-10}\, T^2 + 2.64\ 10^{-10}\, T^3$$

where T is the time or epoch measured in centuries after AD 2000.

the day, the hour and the minute) have failed until now. The most famous example is certainly that of the French revolution at the end of eighteenth century.

2.4. THE SECOND

A definition of the *second*, which was the official definition in the International System of units (SI) until 1960, is the following one: "the second is the 1/86400th part of the mean solar day". The corresponding time scale is the *Universal Time* (UT). It is in fact desirable, for civil purposes, to keep a time-scale which remains in steps with Earth rotation. The universal time is thus defined as the mean solar time of the prime meridian of Greenwich increased by twelve hours.

In the Antiquity, the day and the night were divided in twelve hours, unequal except at the equinoxes or on the equator. The use of these temporary hours remained valid until the 15th century. The astronomers however, even in the Antiquity, already used equinoxial hours of equal duration. Two successive crossings of the meridian by the Sun defined the *real solar day.*

The *mean solar time* (corresponding to a fictitious Sun moving in the ecliptic plane) is the real solar time corrected for some fluctuations which can reach an amplitude of twenty minutes per year (equation of time).

In practice UT is determined from the observation at the meridian of stars whose coordinates are known. Such observations allow to define (with an uncertainty of 0.1 s) a universal time which is referred to as UT_0. In a second step, it is necessary to consider a correction arising from the fluctuations of the North Pole at the surface of the Earth) and to calculate (with an uncertainty of the order of 1 ms) another universal time UT_1, more accurate than UT_0 and which was the basis of the official time scale until 1960. This correction requires however about two months and necessitates the observations of many observatories. Taking into account the seasonal variations of UT_1 related to zonal circulation in the atmosphere allows to define UT_2. In addition, the duration of the mean solar day has been observed to increase by about 10 ms per century. This slowing down of the Earth rotation is due to the Moon attraction and to the energy losses connected to tide effects.

The mean solar time was used by the astronomers but the astronomical ephemeris was expressed in real time. The mean solar time of Greenwich was introduced only in 1834 in the *Nautical Almanach and Astronomical Ephemeris* and, in 1835, the mean solar time of Paris appeared in France in the *Connaissance des temps.*

Figure 1. General view of *Jantar Mantar* (Jaipur Observatory, India). In front of the picture, circular structures (*Ram Yantra*) allowed the determination of the azimut and altitude of celestial bodies. At the back, a large sundial (*Samrat Yantra*), with the hypotenuse indicating the direction of the pole, allowed local time determination with an accuracy of 2 s. In the middle, a set of twelve instruments (*Rahivalaya Yantra*), one for each zodiacal sign, was used for the determination of the celestial latitude and longitude. (Photography by the author)

3. Time measurement in Antiquity

In the Antiquity and until the 15th century, there was little need for an accurate time scale except in astronomy. As a consequence, somewhat "rudimentary" devices were built for time measurement. Let us briefly recall some basic instruments used during many centuries for that purpose.

The *gnomon* is probably the oldest device used for time reckoning and based on the fact that the length and direction of the shadow of a vertical stick are varying during the day. In addition, according to the date considered, for a given hour, the length and direction of the shadow are variable, a variability related to the yearly solar declination. The use of the gnomon is very old and already attested in Ancient China. According to Herodotus, the Babylonians had transmitted it to the Greeks and Anaximander was among the first to use it.

The *scaphe* is a sundial appearing as a hollowed out stone globe containing a pointer at the bottom whose the top does appear at the same level as the edge of the bowl. Twelve graduations, perpendicular to the half-sphere, indicate the daylight hours counted since the dawn. Berosus is probably the inventor of this type of sundial which was used in Ancient Roma.

Originating from Mesopotamia, the *polos* was made of a hollow sphere with the concavity upward oriented. Hanging in the centre of the sphere, a small ball does intercept the Sun light and its shadow is projected on the internal wall where it describes the movement of the Sun. Using this instrument, it is possible to obtain the solsticial and equinoxial dates.

The Egyptians made extensive use of the *water clock* or *clepsydra*. The oldest one is dated from the thirteenth century BC at the time of Amenophis but it could even be older. It was composed of a water tank with a time scale and a hole at the bottom for the flow of water. It is possible that the inventor was Amenembat. Water clocks became rapidly popular in Greece, in the Roman Empire and in the Western countries where they were frequently associated with a sundial. They were replaced during the fourth part of the thirteenth century by mechanical clocks.

4. Instruments used in the Middle Ages

The *sundials* were common during the Middle Ages, the oldest ones being height or altimetric sundials. Based on this principle, the "shepherd clock", known since the 16th century, was providing an approximate estimate of the different hours of the day. Later on, the direction sundials appeared. Between the tenth and fifteenth century many towers and churches were decorated with fixed sundials. Some astronomical observatories have been equipped with large sundials in order to derive a time scale as accurate as possible. A specific example (Jaipur observatory, North India) is illustrated on Fig. 1.

The *astronomical ring* is a universal sundial in copper, brass or silver, made of concentric circles, hanged up by a mobile ring and showing the hour from the capture of the sunlight.

The *quadrant*, already known in Ancient Greece, designates a metal-made instrument limited by two perpendicular sides and by a limb forming a quarter of a circle. A plumb line, fixed at the centre, has a sliding pearl whose the distance to the centre is variable according to the date. When viewing the Sun, it is possible, according to the position of the pearl, to deduce directly the right hour.

Until the end of the seventeenth century, the *nocturlab* did allow to get the correct time during the night from the observation of the pole star by comparison with the pointer stars of the Great Bear (or Dipper). From the

Figure 2. Detail of the astronomical clock of Strasbourg cathedral (France). The present clock is due to J.-B. Schwilgué (1776-1856). (Photography by the author)

measurement of the hour angle of the circumpolar stars, it was possible to derive the hour angle of the Sun and the time.

The *hourglass* or sand-timer is made of two glass bulbs separated by a narrow bottleneck and containing pulverized sand, shell or marble. Some authors claim that the Egyptians already used it.

The planispheric *astrolab*, based on the principle of the stereographic projection developed by Ptolemy during the second century of the Christian era is undoubtedly a master-piece of the Greek geometrical genius. From Greece, it migrated to muslim countries (Byzantium, Syria, ...) after the fall of Alexandria. It became then common in Western countries after its introduction in Spain and in France at the end of twelfth century. The art of the astrolab came to perfection during the sixteenth and seventeeth centuries both in Eastern and in Western countries. Astrolabs were still in use in Persia and in Morocco during the nineteeth century. The late survival of astrolab in muslims countries resulting from its use for the determination of the right times for the ritual prayers according to the Koran.

The astrolab allowed to determine the night and the day hours, the hour of sunset or sunrise, etc ... It played thus, during many centuries, an important role for time measurement and helped much not only the astronomers but also the navigators.

5. Evolution in Modern Times

The inventor of the first clock with mechanical wheels and controlled escapement is unknown. Some claim, but this is unlikely, that Gerbert, a French monk of the tenth century, who became pope as Sylvester the Second (999-1003), was the inventor. A mechanical clock has three essential components: a source of power, a regulator to beat out the ticks and an escapement mechanism.

The first precise descriptions of *mechanical clocks* date back to the beginning of fourteenth century. A famous builder was Giovanni de Dondi, born in 1318, who constructed a planetarium-clock positioned in 1364 in the library of the Pavia castle. Many astronomical clocks were built later on in Europe and among these, one among the most famous was the clock of Strasbourg cathedral in France (Fig. 2).

During the eighteenth century, C. Huygens revolutionalized the watchmaking by contributing to the development of the pendulum and the spiral spring. A significant progress resulted later on from the invention of different types of escapement such as the anchor escapement developed by the English Graham. It is generally assumed that the first watch was due to Peter Heinlein (1480-1542), a locksmith from Nuremberg (Germany). During the eighteenth and nineteenth centuries, the progress in clock development resulted basically from improvements in escapement mechanisms (balance wheel and pendulum) and the errors in time measurement decreased from typically ten to twenty seconds a day to a few hundredths of a second. Timekeeping by navigators was a booster to the development of accurate clocks which culminated with the construction of marine chronometers by J. Harrisson (1693-1776).

The first *electric clocks* did appear around 1840. In such devices, the electricity provides the necessary energy for maintaining the regulator. The electric energy is delivered through a system of piles or accumulators under the form of a continuous electric field. The *electronic clock* uses the miniaturization of the compounds.

The first quartz crystal clock was realized by W.A. Marrison in 1928 and was very accurate: it allowed to reach a precision of 1/1000 s in 24 hours, which was a considerable improvement over the best electric clocks.

The first analogical quartz crystal watch was built in 1967 and the numerical quartz watch was born in 1971.

Naturally, the astronomy directly benefited from all these successive improvements in time measurement.

6. The ephemeris time

It appeared in the middle of the twentieth century that the definition of the second of UT_1 (see Sect. 2.4) had to be refined with respect to its stability. A new time scale was thus proposed in 1956 by the International Committee of Weights and Measures (CIPM). For that purpose, it was decided to use the revolution of the Earth around the Sun as the basis of a new time scale called the *ephemeris time* (ET). This definition was adopted by the eleventh General Conference of Weights and Measures (CGPM) in 1960. According to this definition, which was valid only during the period 1960 to 1967, the second is the fraction 1/31,556,925.9747 of the tropical year 1900 January 0 at 12 h of the ephemeris time. The corresponding time scale was ET. ET was obtained as the solution of the equation which gives the mean geometrical longitude of the Sun according to Newcomb:

$$L = 279°41'48.04" + 129,602,768.13" \ T + 1.089" \ T^2$$

where T est counted in Julian centuries of the ephemeris time *i.e.* of 36525 d. The origin of T was January 0 of 1900 at 12 h ET, when the mean longitude of the Sun was 279°41'48.04.

Theoretically, ET was determined by measuring the position of the Sun in comparison with stars of known coordinates. In practice, such a measurement could not be made directly. The determination of ET was realized by measuring the position of the Moon in comparison with stars of known coordinates, this secondary clock being calibrated in comparison with the displacement in longitude of the Sun. The main difficulty of ET resulted from the fact that a one year wait was necessary to reduce the uncertainty of the measurements (of the order of 0.1 s). On a short time scale, its precision was thus lower than that of UT. One advantage of ET was the long-term stability (about 10^{-9} or 1 s in 10 y).

It should be emphasized that, in view of its somewhat artificial definition, ET was never used for practical life but its use was basically restricted to the astronomers. In 1967, the uncertainties of the available atomic clocks were much better than that of ET (it reached 10^{-12} or 1 s in 30,000 y). As a consequence, it was decided by the thirteenth CGPM to adopt the atomic second as the new official unit of time (see Sect. 8).

7. The local time and the universal time

In astronomy, Noon at a given place, is defined as the moment when the Sun is crossing the meridian. This time clearly is *local* because the meridian of an observer depends upon the longitude. When it is Noon in London, where the longitude is 0°, it is around five hours earlier in New York where the longitude is 74°, the Earth rotating through 15° every hour. The need of dividing the whole Earth in 24 zones of 15°, with some fitting-out for the country borders, arose with the long-distance travels and with the problems created by the train-timetable compilations. In the second part of the nineteenth century, it became necessary to adopt a unique time for a given country: in France, this resulted from the law of March 14, 1891 which imposed the time of Paris meridian. This solution was not adequate for large countries, like the United States, where it became soon necessary to consider several time zones.

By an international meeting, held in Washington in 1884, it was decided that all the countries of the world would establish time zones, the reference zone being centered on the prime meridian at Greenwich and extending from longitudes 7.5° West to 7.5° East. This zone defines the Greenwich Mean Time (GMT). It was decided also that the day of UT would begin at Midnight, in disagreement with the common practice of the astronomers to start the day at Noon. These were reluctant to adopt the new system: the GMT (but not the UT which is differing by 12 hours) as defined in 1884 was used only in 1916 in *La Connaissance des temps* and UT (beginning of the day at Midnight) was adopted only in 1925 in the ephemeris for navigators and astronomers. The new time scale, obtained from the mean time by adding 12 hours, was defined as the *civil time*. The use of UT has been strongly recommended by the IAU in 1948.

The realization of UT, the time based on Earth rotation and counted from the prime meridian, was not evident. From the middle of the nineteenth century up to around 1970, the astronomers observed stars in the meridian plane and noted carefully their passage with the help of accurate clocks. Some devices emitting hourly signals were also in use. The observations were then reduced in order to derive the corresponding universal time. With the advent of the radio signals around 1910, it appeared that the systematic errors (of the order of the second) could be larger than the uncertainties affecting the local time scales (of the order of the ms). This gave rise to the Bureau International de l'Heure (BIH) which became operational in 1919 (until 1988[3]) and whose the main aim was to produce a unique approximation of UT called the *definitive hour* and to provide the observers with accurate time intervals between the definitive hour and the times of emission of the hourly signals.

[3]In 1988, the BIH was devided in two parts: the *Bureau International des Poids et Mesures* (BIPM) is in charge of the atomic time and the *Service International de la Rotation Terrestre* (IERS) is in charge of astronomical and geophysical activities.

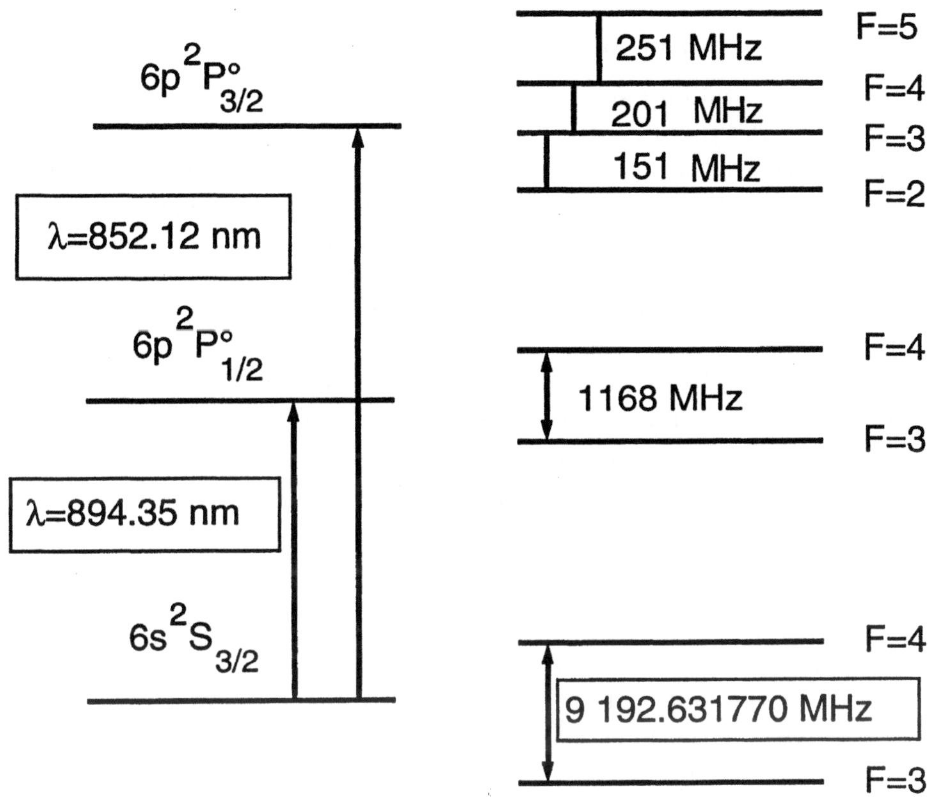

Figure 3. Partial energy level scheme of the caesium atom showing the transition considered for the definition of the second.

8. The atomic time

Since 1967, the second, according to the CGPM, is defined as: "the duration of 9,192,631,770 periods of the radiation corresponding to the transition between the two hyperfine levels of the ground state of Caesium-133 atom" (Fig. 3). The adoption of this definition, the unit of time of the International System of units (SI), has open the door to the atomic time era.

The corresponding time scale is the *International Atomic Time* (TAI). TAI is established by the Bureau International de l'Heure (BIH) (now replaced by the Bureau International des Poids et Mesures, BIPM) on the basis of atomic clocks in use in different places through the world.

TAI is now the official time for dating the events and is considered as the most accurate time scale presently available. It is possible to evaluate the characteristics of this time scale by considering:

- the time interval between TAI and each clock k participating to its definition TA(k);
- the uncertainties of connection.

In 1998, the stability and exactness of TAI were estimated to be 2.10^{-14} (1 s for 1,500,000 y).

To construct an atomic clock, one has to use the frequency of the radiation emitted by the atoms and to count the periods of an electromagnetic wave producing the change of state of the atoms (passive standards) or generated by this change of state (active standards).

There exists presently different types of atomic clocks: the caesium and the rubidium clocks (more limited capabilities), the passive hydrogen masers and the active hydrogen masers whose short-term stability is better than the caesium standard but whose long-term stability is lower.

In a caesium atomic clock, a quartz oscillator or a hydrogen maser coupled to an electronic device is used to produce an oscillating magnetic field with a frequency of 9,192,631,770 Hz and this hyperfrequence signal is injected into a waveguide which maintains a resonance at this specific frequency (Ramsey cavity). For maintaining the frequency with accuracy, a beam of caesium-133 atoms in different energy states E1, E2, ... is produced by a furnace and there is a selection, by magnetic deflection, of the atoms populated in the state E1, those atoms only being allowed to enter the Ramsey cavity. If the injected frequency is exactly 9,192,631,770 Hz, one can observe transitions of many atoms from the state E1 to the state E2. Those atoms, populated in state E2, are separated by a second selection system from the atoms populated in state E1 and detected. According to the response of the detector, there is a modification of the quartz frequency in order to optimize the detection of the atoms in state E2. It is thus a quartz crystal oscillator which is the starting point of a caesium atomic clock, the atoms of caesium being present to control and adjust the frequency of the signal generated by the quartz : it is a passive standard.

Concerning the atomic time scale characteristics, the following considerations apply:

- the time scale is an integrated scale *i.e.* it is realized by a superposition of time intervals resulting from a standard;
- no variation in the stability has been observed at the level of 10^{-14} yr^{-1};
- the duration of the second used in the construction of an atomic time scale must agree as strictly as possible with the SI definition. At the turn of the century, this agreement was about 10^{-14} and it is anticipated that a further improvement of 1-2 orders of magnitude could still be gained;

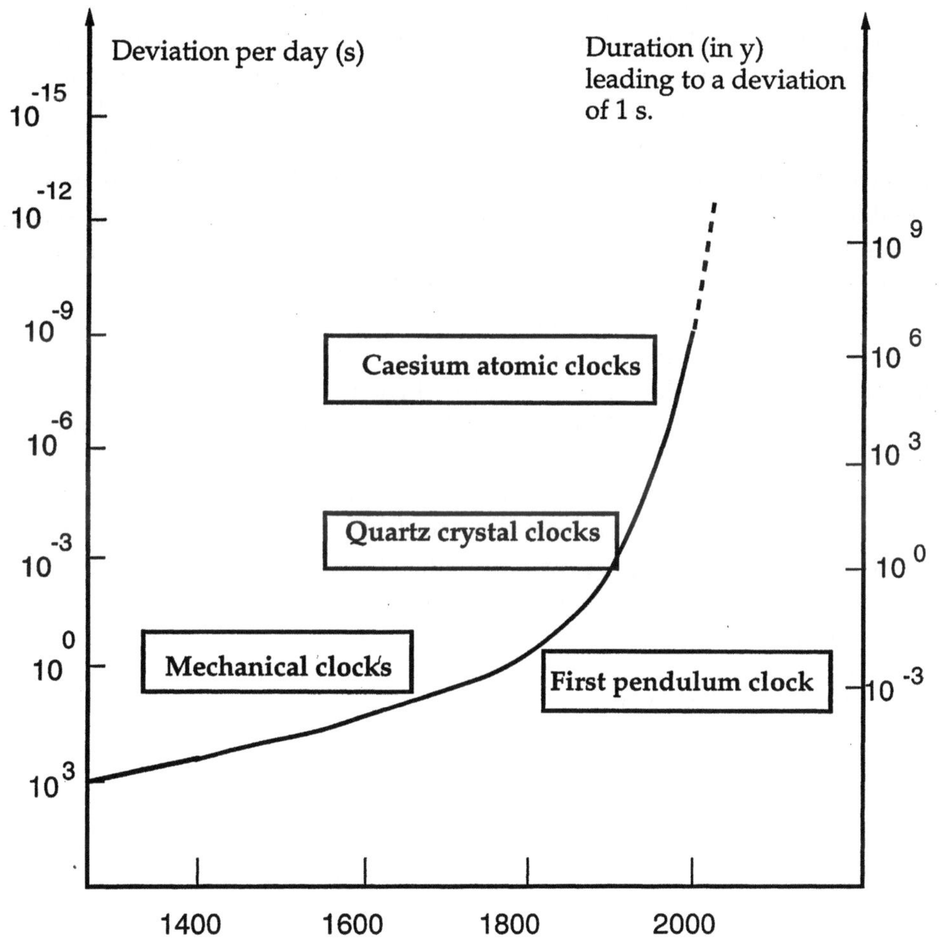

Figure 4. Evolution of the quality of the clocks as a function of time for the period 1300-2000 AD.

- TAI is no more a natural time scale but, for defining it, it is necessary to rely on man-made clocks. The atomic time scale of an isolated clock is not everlasting and perennial and this lack of perenniality is corrected in TAI by the link of several clocks;
- the universality is guaranteed by international organisms (IAU, CIPM, ITU-R, ...);
- the unit of time is now very accurately known (a few 10^{-14}).

A difficulty when using TAI results from the progressive shift of TAI compared to UT. TAI must remain in phase with UT which is directly related to Earth rotation and is important for human life. As a consequence, it has been decided to define the *Coordinated Universal Time* (UTC) used for generating the legal time of all the countries and directly related to TAI, the largest difference when compared to UT_1 reaching 0.9 s. This is made possible by adding, from time to time, a leap second to UTC. One second of this type has been added *e.g.* on December 31 1998 at 23 h 59 m 59 s and on January the first 1999 at 0 h 0 m 0 s.

For obtaining TAI, the laboratories involved in the process must have several atomic standards for comparing the different scales of local atomic time and know the delay or the advance of the local scale when comparing it with other laboratories. TAI is obtained as the weighed mean of the different local times. Each laboratory receives then the correspondence between its local scale and TAI for some events of a given period which provides the opportunity to redate them according to TAI.

Presently, these comparisons are realized through the use of the Global Positioning System (GPS) with an exactness of the order of 3 ns.

Some characteristics of the atomic time are compared on Fig. 4 to previous scales on a curve showing the evolution of the precision of clocks for the period 1300-2000 AD.

9. The Julian date

The Julian period defines a cycle of 7980 years familiar to the astronomers but is not related to the chronology associated to the Julian[4] calendar. This period consists in a continuous series of days (excluding thus the weeks and the months) starting at Noon on January 1, 4713 BC. It is generally considered that it was first proposed by the French Joseph Julius Scaliger in his work, *Opus de emendatione temporum*, published in 1583 (Fig. 5). The Julian period is a combination of three cycles of 28, 19 and 15 years leading to $28\times 19\times 15 = 7,980$ years. These cycles correspond respectively to the solar cycle (28 years), the lunar cycle or Meton cycle (19 years), well

[4]From the name of Julius Caesar who reformed the Roman calendar.

Figure 5. Drawing of the French scholar J.J. Scaliger. (*in* Isaac Bullart, Académie des Sciences et des Arts, Brussels, 1695 (© Library of Liège University, Belgium).

known since the Greek Antiquity, and the Roman cycle of the indiction (15 years) used for tax payment. These three cycles have different origins: 9 BC for the solar cycle, 1 BC for the lunar cycle and 3 BC for the indiction cycle and, for getting a common origin for the three cycles, the year 4713 was adopted.

This period is frequently used in astronomy, each event being identified by a number of days numerically counted since January 1, 4713 BC at 12 h., the Julian days (JD) starting at 12 h (UT). The Julian day starts 12 hours after the Greenwich Midnight at which the corresponding day starts. As an example, January 1, 2000 at 6 pm GMT corresponds to the day 2,451,545.25 day of the Julian period. The decimalization of the days allows an easy deduction of a time interval between two astronomical dates *e.g.* for the study of the variable stars. The day which starts at Noon on January 1 and which ends on January 2, at Noon is counted as January 0. In that way, the day starting at Noon on January 0, 1900 (December 31,

1899) is numbered 2,415,020.0.

The astronomers frequently consider the modified Julian day (MJD) which can be deduced from the number of days of the Julian period by subtracting 2,400,000.5. This procedure corresponds to the adoption of a time origin on November 17, 1858 at 0 h (TU) and has been officially adopted by the IAU in 1973.

10. The coming back of the stars

New, very accurate celestial clocks have been discovered in 1967 thanks to the radiotelescope of Cambridge (UK) by J. Bell and A. Hewish. This discovery is looking like a hopeless attempt of the astronomers to acquire again the control of time measurement that they were just loosing! These new clocks are pulsars of which more than 1000 were identified in 2000 AD, most in our own galaxy[5] and with pulse periods ranging from 1.557 ms to over 8 s. The astronomers distinguish the ordinary pulsars and the binary and millisecond pulsars.

Only a small number of supernova remnants harbour radio pulsars. The well known association between the Crab Nebula (observed in 1054 AD by Sung Chinese astronomers) and the pulsar PS B0531 + 21 strongly emitting in the radiofrequency domain but also in the optical and in the short wavelength ranges (X-rays and γ-rays) is particularly important and has given to the astronomers a precise age for the pulsar.

About 50 pulsars are known in binary systems, the first of which was discovered by R. Hulse and J. Taylor[6], in 1974, during a survey for new pulsars carried out at Arecibo observatory. Pulsars are, in view of their regular clockwise pulses, sensitive probes of the gravitational environments in which they are found.

The periodic character of the pulsar emission is explained by a rotation phenomenon: pulsars are rapidly spinning neutron stars resulting from the explosion of supernovae that accompany the collapse of massive stars. Their diameter is no more than a few tens of kilometers but the associated effects are spectacular:

1) due to the conservation of the kinetic angular momentum, the pulsar rotates rapidly on itself (the period being of the order of the second for the "standard" pulsar and of the order of the millisecond for the "millisecond" pulsars)

2) The magnetic field, confined in a limited volume, is very intense and produces conical beams of electromagnetic radiation that sweep past the Earth producing pulses primarily observed at radio wavelengths.

[5] A few of them have been detected in Magellanic Clouds.

[6] They have been awarded the 1993 Nobel Prize of Physics.

In fact, two beams of radio waves are emitted along the magnetic axis of the star which generates electromagnetic radiation in the way of a rotating lighthouse. As the magnetic poles of the star do not generally coincide with the geographic poles, at each rotation, if the beam is directed toward the Earth, the observer receives a pulse of radio waves. The ordinary pulsars emit during a few tens or a few hundreds of millions of years and then switch off. The periods of the pulsars, although remarkably constant, slightly increase with time due to the slowing down of their rotation. In the case of the Crab Nebula, the increase reaches 10 microseconds per year.

After a detailed investigation of the impulsions received from the first millisecond pulsar (PSR 1937+21, discovered in 1982), it has appeared at the end of the eighties that these celestial bodies could satisfy the stability criteria for defining a new astronomical time scale. However, although the emission of electromagnetic waves by these somewhat strange objects is characterized by a surprising accuracy, this one is still smaller than that reached by the atomic clocks.

Due to the low signal-to-noise ratio of the millisecond pulsars, their short-term stability is limited. For example, in the case of PSR B1937+21, the first millisecond pulsar discovered, the period is equal to 1.557 ms and it is only possible to know the impulsions with a precision of 0.5 ms, which means a stability of 3.10^{-4} only. The long-term stability however is much better because, in the case of the latter pulsar, the astrophysicists were able to register the impulses since the time it was discovered.

11. Spatial projects and future evolution

It is planned that the International Space Station (ISS) will welcome from June 2005 an atomic clock which will be able to measure the second with an unprecedented accuracy of 10^{-16}, the lack of gravity giving the opportunity to refine time measurement by a factor of about 5. This experiment, called PHARAO[7] will be conceived and financed by CNES. It will be the masterpiece of Project ACES (Atomic Clock Ensemble in Space), proposed by ESA, and including a hydrogen maser with a connection to Earth by laser and by microwaves. The main purpose of the experiment will be to compare the frequency emitted by an atomic clock on the ground (Paris Observatory) with the frequency emitted by the clock at an altitude of 400 km. The comparison of the frequencies will allow, *in fine*, to test Einstein's theory of general relativity.

One of the major causes of instability of the atomic clocks is due to the thermal movement of the atoms. This results in a frequency modification

[7]From "Projet d'Horloge Atomique par Refroidissement d'Atomes en Orbite" in French.

related to the Doppler effect. The ideal would consist in using atoms with a temperature close to the absolute zero (0° Kelvin = -273° Celsius). It is therefore necessary to cool down, by laser radiation, the atoms after their ejection from the furnace by heating. If an atom, enlightened by a laser, whose the frequency is slightly lower than a typical frequency of the atom. is motionless, it does not interact with the beam and, consequently, remains motionless. If, on the other hand, it comes closer to the laser light, it will see the frequency of the beam as equal to its own transition frequency (Doppler effect). It absorbs a photon and will slow down. When adding a second laser in the opposite direction, another couple of lasers in the horizontal direction perpendicular to the first one (one in each direction) and finally an additional couple of lasers in the vertical direction, on gets "optical molasses" and the atoms are slow down independently of their direction of propagation. It is now possible to cool down the atoms to a temperature of 2 micro-Kelvin. A clock with a fountain of caesium atoms cooled down by laser is working at Paris Observatory. The stability of such a clock is estimated at about 5.10^{-16} on a day.

Accurate atomic clocks in space are very useful for positioning systems like the GLONASS Russian system and the GPS developed by the USA allowing to deduce immediately a position with about 20 meters of error. For the future European positioning network GALILEOSAT, which must be operational in 2007, hydrogen masers will be used allowing to deduce a position with an error of 30 centimeters. The clocks used in these systems however will not be cooled down.

12. Conclusions

To the vague cutting off of the daylight in Antiquity, the increasing technical nature of the evolving societies has imposed a more and more constraining fragmentation of the days to generate *in fine* smaller and smaller parts of the second. The laboratories now measure currently radiative lifetimes reaching the nanosecond or the picosecond and, with modern laser devices, it is possible to reach the femtosecond or even smaller time intervals. On the other side of the time scale, dating techniques used in geology, paleontology or astrophysics, (*e.g.* the disintegration of uranium in lead) now allow a scientific estimate of very long time durations like the age of the Earth.

Since ever, mankind has been dependent upon day and night rhythms associated with Earth rotation. During many centuries the constraints concerning the timekeepers were not very severe: this allowed to use and to improve the antique water clocks, the sand- timers, the sumptuous astrolabs and the monumental clocks. With the advent of industrialization, more accuracy and precision were needed. When the day variations became obvious, the definition of a mean solar day by the astronomers was necessary and a first definition of the second was proposed as a conventional part of the day. In the middle of the 20th century, the requirements of stability, accuracy and precision of the second (reaching relative accuracies of about 10^{-7}-10^{-8} in the astronomical definition) were no more sufficient for scientific (radar ranging, spectroscopy) or for technical (telecommunication, electronic instrumentation) applications. In 1955, a frequency standard, based on a caesium transition, was put in operation in England and, with its accuracy of 10^{-9}, allowed to measure the decrement in the speed of Earth's rotation. With the advent of the atomic clocks, we could observe a real revolution in time reckoning and a TAI unit, adopted in 1967, could be defined with an unprecedented accuracy. It has also opened a new era for astrophysics because the time control has been definitely lost by the astronomers to the benefit of the atomic physicists ...

References

1. Audoin, C. & Guinot, B. 1998, Les Fondements de la Mesure du Temps, Masson, Paris.
2. Biémont, E. 2000, Rythmes du temps, Astronomie et Calendriers, De Boeck Université, Paris – Bruxelles.
3. Fraser, J.T. 1987, Time, The Familiar Stranger, Tempus Books, Washington.
4. Lippincott, K. 1999, The Story of Time, Merrell Holberton Publ., London.
5. Lyne, A.G. & Smith, F.G. 1998, Pulsar Astronomy (2nd Ed.), Cambridge University Press.
6. Manchester, R.N. & Taylor, J.H. 1977, Pulsars, San Francisco, CA, Freeman.
7. Murdin, P. (Ed.) 2001, Encyclopedia of Astronomy and Astrophysics, Institute of Physics Publ., London.
8. Richards, E.G. 1998, Mapping Time, The Calendar and its History, Oxford Univ. Press.

Common abbreviations
(asterisks indicate international abbreviations)

ACES	Atomic Clock Ensemble in Space
BIH	Bureau International de l'Heure (France)
BIPM	Bureau International des Poids et Mesures (France)
CGPM	General Conference of Weights and Measures or, in French: Conférence Générale des Poids et Mesures
CIPM	International Committee of Weights and Measures or, in French: Comité International des Poids et Mesures
CNES	Centre National d'Études Spatiales (France)
ET	Ephemeris Time
GLONASS	GLObal NAvigation Satellite System
GMT	Greenwich Mean Time
GPS	Global Positioning System
IAU	International Astronomical Union
IERS	International Earth Rotation Service
ITU-R	International Telecommunication Union – Radiocommunications
ISS	International Space Station
JD	Julian Date
MJD	Modified Julian Date
SI*	International System or, in French: Système International
TAI*	International Atomic Time or, in French: Temps Atomique International
TA(k)*	Atomic Time defined by the laboratory k or, in French: Temps Atomique défini par le laboratoire k
TT*	Terrestrial Time
UT	Universal Time
UTC*	Coordinated Universal Time
UT_0*	Universal Time, Form 0
UT_1*	Universal Time, Form 1
UT_2*	Universal Time, Form 2

EVOLUTION OF DATA PROCESSING IN OPTICAL ASTRONOMY – A PERSONAL ACCOUNT

R. ALBRECHT
Space Telescope European Coordinating Facility†
European Southern Observatory
Karl Schwarzschild Straße 2
D-85748 Garching, Germany
ralbrech@eso.org

Abstract. This paper covers more than 30 years of development of data processing in optical astronomy. This is not a review paper, but rather an account of the history as seen by somebody who has been involved hands-on in building data analysis systems at different institutes, for different astronomical instruments and for different generations of computers. It is, by necessity, a very subjective account.

1. Introduction

As I am writing this in mid-2002, trying to remember some of the details of the developments which happened about 30 years ago, I am using a computer which not only accepts and formats the text which I am entering, but also plays, at the same time, Bach's Wohltemperiertes Klavier from an mp3 file and accepts incoming email while a webcast is being displayed in a browser window. All this is happening without pushing the limits of the machine. We have come a long way.

Today astronomy and computers are synonymous. But this was far from clear at Vienna Observatory in 1967. Having just started a thesis in the very traditional and then very conservative field of astrometry, I was faced with the fact that measuring the plates on a manually-operated measuring machine would take me the better part of one year. And that reducing the

† Affiliated to the Astrophysics Division, Space Science Department, European Space Agency

A. Heck (ed.), Information Handling in Astronomy – Historical Vistas, 35-60.

measurements using the then undisputed logarithmic method was going to take another year. This prospect was not exactly appealing. I decided to not do logarithms and instead invest the year in learning to program computers.

This was not as easy as it sounds. In 1967 hardly any observatory had a computer, and Vienna Observatory certainly did not have one. All we had was mechanical calculators. Some of them could even perform divisions!

In fact, not even Vienna University had a computer. In 1965, at the occasion of the 600th anniversary of the University, IBM had offered a computer, the newly developed IBM 360, at an advantageous price. But the computer had a long delivery time, and anyway, the rooms had to be adapted first. The IBM 360 finally went into operation in early 1968.

The Technical University in Vienna did have a computer. It was a battleship gray IBM 7040, complete with "Blinkenlights" – some of the readers will remember the pseudo-German sign, which appeared everywhere – and a staff of white-coated technicians who were busy feeding the beast. With power, lots of it, and with punched cards, tons of them. And so I took computer courses at the Technical University. The terms *Computer Science, Informatics*, and *IT* had not been invented. The field was modestly called *Numerical Computing Technology.*

A Fortran IV compiler had just been installed, so this was the new bandwagon. Although the purist faction insisted that Algol, being structured, was a much better language, the engineers, being pragmatists, went with Fortran. `COMMON` and `DATA` statements appeared to be the elegant solutions to many problems. And so it was Fortran IV on the IBM 7040 with which I did my first astrometric plate solutions.

When the IBM 360 became available, I took my code to the new machine. After all, it was faster, more modern, had almost no users at this point, and was located much closer to the Observatory. And I quickly found what I was going to find over and over again in the future: compatibility was a myth. Fortran IV on the IBM 7040 was not Fortran IV on the IBM 360. Worse than that – even after I had gotten the code to run on the IBM 360, which, among other nuisances, meant coming to grips with the infamous IBM Job Control Language (JCL), I found that the results differed. Not by much, to be sure, the culprit was the difference in word length between the two machines. But this taught me some basic lessons for the future.

It also opened my work to criticism: clearly this would not have happened with the logarithmic method. All right, so I claimed that I had found the reason for the discrepancy and that I had eliminated it. But was this true? And: how can you trust a number if you did not "compute it yourself"? It was seriously suggested to me to take my thesis to the numerical computing technicians. I refused. Not only because I resisted the career change, but also because by now I firmly believed that the computer had a future in astronomy. It was only after I started to do Monte-Carlo type experiments when the conservatives finally yielded: checking what happens to the accuracy of plate solutions if one, or several, and different, standard stars were omitted? How does the distribution of standard stars change the results? What happens if one "rattles" the standard stars in their error boxes? This required hundreds of computing runs per plate, something, which was clearly impossible using the logarithmic method.

As it turned out, it was not too easy on the 67-vintage IBM 360 either. It took hours and hours of computer time. But I was one of no more than ten serious users of the machine at that point and the time was not charged for. So it was possible. In 1969 we ran the machine into the ground. It was a hot summer and the air conditioner could not cope, so the IBM 360 came to a grinding halt. The administrators discovered that for the last several months the machine had been the private toy of a very few people, so they shut it down and kicked us out. The facility was re-opened in Fall as one of the infamous university computer centers, where your interface to the computer consisted of a pigeonhole into which you stuck your punched cards and in which you found your printout 24 hours later. The fun days were over. But my work was completed. And a passion for interactive, hands-on work with computers had started.

2. The Beginnings

1968 was also the time when Karl Rakos came back to Austria after several years in the United States. To him it was obvious that astronomy without computers was impossible. Not only for numerical calculations, but even more so for controlling instruments and telescopes. And so in 1969 Karl initiated the procurement of a Digital Equipment Corporation PDP-12, one of the stranger machines in the zoo of what was then called mini-computers.

The PDP-12 was a combination of a PDP-LINK and a PDP-8. The former was designed to be a laboratory machine. It had things like controllable relays for switching machinery, potentiometers and analogue inputs, and even an oscilloscope-type screen to visualize signals. The PDP-8 was a more modern design, was very popular, and there was a lot of software for it. The PDP-12 was DECs attempt to introduce the PDP-8 into the laboratories by allowing to run LINK code on it. The machine consequently had two different processors and two different instruction sets. The two processors shared 8K of memory. I/O was only possible in LINK mode, computing was better done in 8-mode. The operator interface was an ASR-33 Teletype, then de rigueur. Together with Helmut Jenkner we went into production reducing spatially-resolved (area-scanner) photometry data.

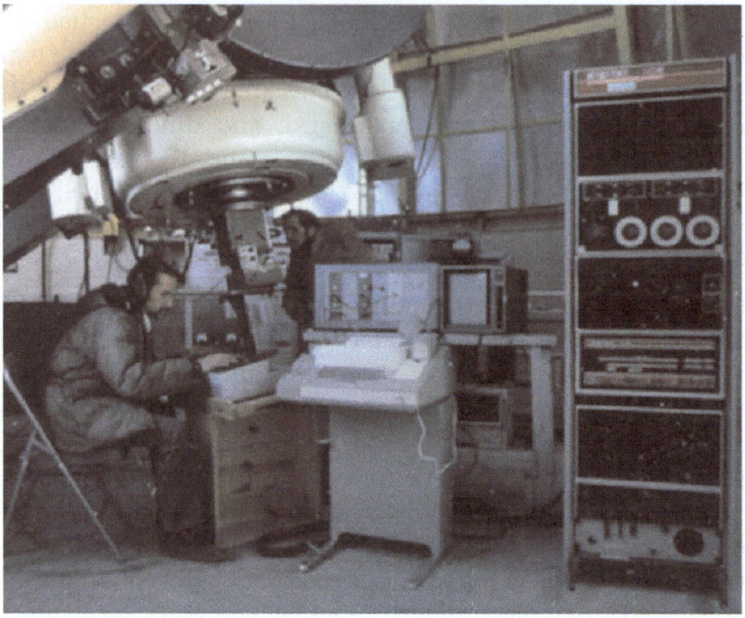

Figure 1. The Lowell Observatory Data System at the Lowell 24 inch telescope. Peter Boyce (at the telescope) and the author during one of the first tests in 1971. (Photograph by the author)

I had come in (academic) contact with assembly language programming through my courses at the Technical University. Now I really got into it, often using the switch register to make modifications to programs. Squeezing instructions into the 200 (octal) words of a memory page became a fine art. We also patched the magtape (Link-tape) driver to allow direct addressing of tape blocks, swapping memory loads to and from magtape in seconds – the first appearance of virtual memory. Why did we not do this on the hard disk? Because the machine did not have one.

The PDP-12 with its oscilloscope would become the first machine on which I displayed an image: by turning off all lights and slowly painting 256 lines consisting of 256 points each, line after line, while a camera with an open shutter looked at the screen. The data were read from magtape, building up the image took several minutes. Stray light from the blinkenlights was a big problem.

3. Early Machines

In 1970 DEC introduced the PDP-11. Peter Boyce at Lowell Observatory in Flagstaff, Arizona, had decided that the time had come to computerize the instrumentation of the main Lowell telescopes, and that the new PDP-11

Figure 2. Peter Boyce (right) and Jim Chastain installing the Lowell Observatory Data System, stripped to its essentials, on the 88-inch telescope on Mauna Kea, to drive an area scanner for observations of Mars. (Photograph by the author)

was just the machine to do it with. Peter was looking for someone to help him do this, so I went to work for him.

We started with PDP-11 Serial Number 177, a paper-tape-based editor and Version 1.0 of the assembler, also on paper tape. In addition to that, we had a soldering iron and other assorted tools, plus enormous confidence to match our ignorance. Which was just fine, hardly anybody else knew any more than we did. Within two years, we had developed our own little magtape-(DECtape)-based operating system, which included an interactive command interpreter, application software for the real-time control of spectrographs, area scanners and high-speed photometers, and analysis software written in BASIC. No image processing, we only had one-dimensional data.

Modest as this effort might appear as seen from today, it was a major paradigm shift at the time. Our system was based on the principle that the hardware could no longer work independently without the computer. We had no dedicated hardware to scan our spectrum. Instead the photon counter was read by the machine and displayed on a Tektronix storage scope for which we had developed graphics software. We had replaced the ASR-33 Teletype by a solid-state keyboard to work in the cold environment of the dome. The command for each step of the stepping motor came from the computer through interface units, which we had built ourselves. This foreshadowed today's embedded processors, but was considered heretical at the time.

There was an early contact with serious image processing. We were observing on Kitt Peak, and we used the occasion to visit Don Wells who was working in Tucson. He showed us the Image Processing System (IPS), which he was in the process of building up. We were impressed by the computer and by the image display hardware, both very powerful but very expensive, and we concluded that image processing would forever be outside our financial capabilities.

The Lowell Observatory Data System, as we called our brainchild, kept working long after Peter and I had left the observatory. In fact it was still working even at a time when computers were cheaper and more powerful – the "dumb but faithful servant" did its job and did it well. It was finally retired in 1983 when the PDP-11 Serial Number 177 had become an anachronism and maintaining it became impossible.

4. ESO Chile

Spectrum scanners were popular in the seventies. By 1973 John Wood had built a scanner at the European Southern Observatory in Chile. It was in dire need of being computerized. Given my previous experience at Lowell, the job was rather straightforward, except that the computer this time was a Hewlett Packard 2114B and the graphics I/O device was a surplus storage oscilloscope with a 10 by 15 cm screen which Walter Nees soldered into the machine. We operated it by gutting and patching the HP plotter driver software and combining it with PDP-11-derived character generator and vector display software. The virtual memory concept was adapted, this time properly as virtual arrays, being swapped between memory and hard disk. And the analysis software was developed along the lines of the successful implementation at Lowell. Listo!

Jim Rickard, also at ESO, had procured a SIT Vidicon camera tube, which he intended to use for first experiments in two-dimensional imaging. We rigged a test setup in the laboratory and pretty soon we had the HP 2114 talking to the camera and displaying test patterns at first and then real – dithered – green-and-white images on the tiny screen. Now this was cool (or, rather, groovy, as we used to say in those days) – our very own poor man's version of the IPS.

Figure 3. The author at the ESO General Purpose Data Acquisition System in 1973. It was used to drive the spectrum scanner in the Nasmith focus of the 1.5 m telescope on La Silla. (Photograph by H. John Wood)

Interactive instrument control software and various quick-look data analysis routines followed naturally and rapidly – all within less than three months. Unfortunately, however, both the SIT Vidicon and the spectrum scanner, along with their originators, fell victim to the closure of ESO operations in Santiago in the mid-seventies.

5. Cerro Tololo/Vienna

Cerro Tololo Interamerican Observatory was (and still is) the friendly competitor of ESO in Chile. I had been visiting and seen the work of Barry Lasker, who did, with Data General Nova computers, things which were very similar to what Peter Boyce and I had done at Lowell. Data acquisition and data analysis being separate issues, a computer center had been established at the La Serena office of CTIO. This one-man computer department was operated by Skip Schaller. And Skip had already started to build an image and spectral processing system which he called TV[1] System, but which would later be known as the Tololo-Vienna System.

The Harris Datacraft computer at the Tololo office was, by comparison to the DECs and HPs that I had previously worked with, extremely powerful. In addition it had several 6-Megabyte disk drives (the size of trash compactors). Barry Lasker, having realized that imaging was going to be the next big thing in optical astronomy, had made sure that we would be able to accommodate all these pixels. And it was in this comfortable computing environment, a powerful (for 1976) machine, plus the beginnings of a command-line-oriented data analysis host system that we started to do serious image processing. Previous concepts, like virtual arrays and display routines, were re-used, but this time in Fortran rather than in assembly language. Within a very short time, a large body of image and spectral analysis programs had been built up.

Images and spectra were also collected at Vienna Observatory. We had procured one of the first, if not the first, PDS scanner in astronomy. Obviously the by-now-aging PDP-12 was totally inadequate to analyze data which the PDS produced in huge quantities. Following the bandwagon of the time, a PDP-11 had been procured. That machine had matured since the days we used it at Lowell – a vendor supported operating system and high-level languages were now available. And I had an image processing system, written in Fortran, which could be used to analyze the PDS data. Except, again, and this time less surprisingly, I quickly found that Fortran on the Harris Datacraft was not Fortran on the DEC PDP-11.

And so the desire for portable code, hardly a worthwhile consideration during the era of assembly language programming, was not born out of academic considerations of how things should be done in an ideal world, but out of sheer frustration. I had a massive amount of powerful software on a magtape, but it just did not compile. And if I changed it such that it would, any software written for the PDP-11 in Vienna was not going to compile at Tololo.

Coordination with Skip Schaller was extremely difficult. No email, no faxes in 1977. No direct dial phone calls, even. In fact, the Chilean telephone system outside large cities was still based on manual switchboards and human operators. There was telex traffic via the CTIO office in Tucson, which operated a shortwave service to La Serena. In an emergency, this channel could also be patched into the US phone system and used for voice communication. Crackling, fading, unidirectional, with two operators at either end for direction switching. Over. Say again. Over. No good. Over. At least not for useful discussions. Over and out.

[1]TV for TeleVision, after the black-and-white TV set which was used for image display.

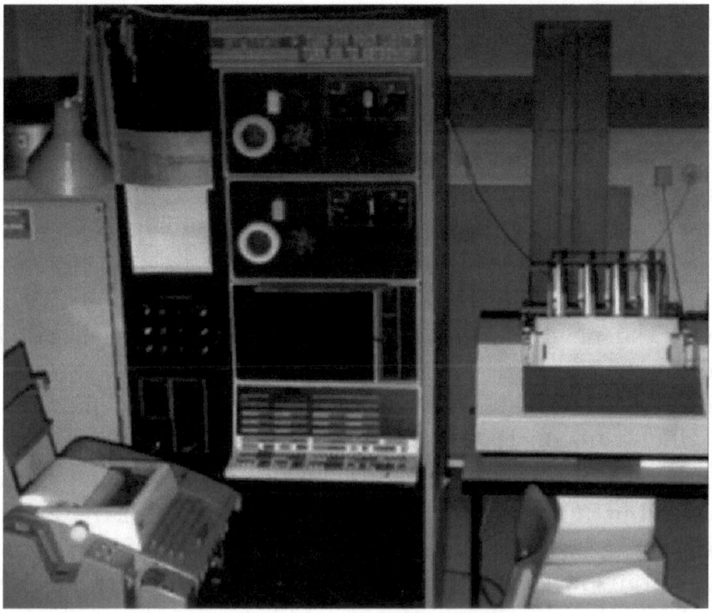

Figure 4. The Digital Equipment Corporation PDP-12 Computer at Vienna Observatory in 1975. It had 8K of magnetic core storage and a 64K hard disk. (Photograph by the author)

So coordinating happened during visits, which means it took considerable time and did not produce a perfect result. However, it did become possible to run application software of the TV System on both computer installations with a modest amount of effort.

The next problem came from an entirely different direction. The system had become very popular and people started to contribute software to it. While most contributors respected the rules and conventions, which Skip and I had agreed on, some of the more ambitious contributors happily, and with the best of intentions, began to introduce their own. Things started to become extremely serious when people undertook to make changes to existing software without following proper procedures. Quite often this meant that scripts (of course our command interpreter had a scripting language) failed to work, or, worse, programs produced wrong results. The spectre of configuration control had raised its ugly head.

And the issue of testing as well. People would write or modify software, use it successfully for their data sets, and contribute it to the system. More often than not the software would fail when used on somebody else's data sets.

In short, we were quickly discovering the very same problems which all other producers of software were discovering during this period of time, and which lead, in many cases, to unsuccessful software projects. The result was a flood of software project management publications in the early eighties, most of them wrong, as people tried to find ways of this situation. We pulled through, not because our software development methodology was superior, but because the authors of the software were singularly stubborn and they were interested in the ultimate goal of the whole activity: finding out how the universe works.

By 1978 the system was in use at several observatories, among them Lowell and the Carnegie Department of Terrestrial Magnetism (DTM). STARLINK, the newly formed UK astronomy network showed interest, but they did not have PDP-11s. They had the new thing, the VAX-11/780. Not to worry, we thought. The VAX was supposed to effortlessly run all software, which runs on PDP-11s. Or so we thought.

Well, not only were the Fortran compilers different, hardly a surprise at this stage, but all the intertask communication failed to work. About the only advantage was the increase in speed and the fact that we did not have to write our own virtual array handlers any longer: the VAX, which stands for Virtual Array Executive, provided virtual arrays. Imagine our surprise when we found that we could not use them! Our convention was that images were worked on in a row-by-row fashion, which the author of any new software had better kept in mind, in the interest of minimizing I/O time. The VAX VMS operating system did not know this and made its own decisions about which subset of an array it would swap. In the worst case, this had the effect that the operating system swapped memory loads every time the application program addressed a new pixel, even if it was an adjacent pixel on the same image line. We had to fall back to our own image I/O.

The TV System in its various incarnations was in use at different institutes and observatories until about the mid-eighties, coincidental with phasing out the VAXes. There was good collaboration among the contributors. Informal workshops were held and a newsletter was circulated. The newsletter eventually turned into the "Astronomical Image Processing Circular" and a total of 10 issues appeared on a semi-regular schedule. It had articles, but it also had printouts of working code (no CD-ROMs at the time), which people could key into their terminals. This software was eventually collected onto a magtape and made available to the readers of the Circular.

The TV System was a successful attempt at bottom-up software coordination with a minimum of imposed structure and essentially no dedicated resources. In terms of stick-and-carrot, it worked with only the carrot (the software that it contained), and even the carrot did not come all that easy. It is difficult to say how much it contributed to astronomy. It did save

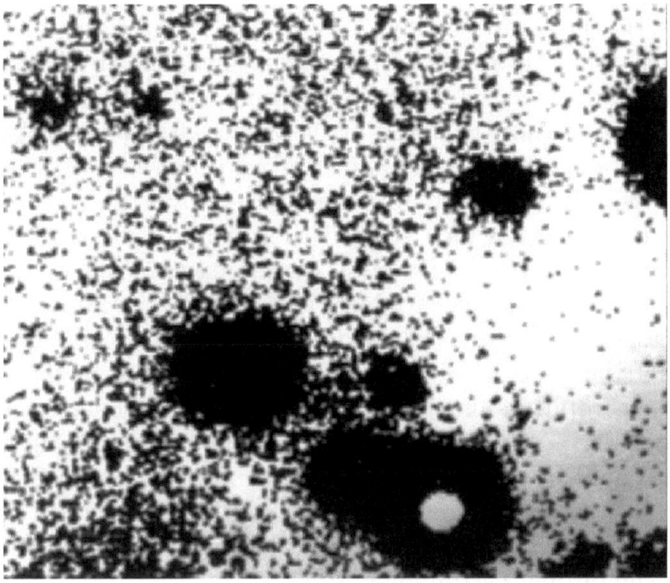

Figure 5. Early image processing. A stellar field displayed line-by-line on the screen of the Vienna Observatory PDP-12. The image was captured by a camera with an open shutter. It took about 30 seconds to record the image. Note the wraparound in the core of the brightest object (1975). (Photograph by the author)

considerable amounts of software development work, and it made people aware of the issues related to the sharing of software. It worked as well as could be expected, and certainly much better than some other efforts. It was carried by the enthusiasm and the dedication of the contributors, for which they deserve to be commended.

6. Portability

Having developed software in assembly language, the antithesis of portability, and thus having had to re-write software from scratch whenever I took it to another machine, and having seen the advantages of working with a data analysis host system which was able to accept software from many different contributors, the concept of software portability had become self-evident to me and to everybody else who was involved in the TV System. A wide community, working on different hardware platforms and vendor-supplied operating systems was using software, which other people had developed, and was developing software, which other people could effortlessly use. This amounted to a multiplication of resources, which was very important for the small institutions, which used the TV System.

There was another quite tangible and economically quantifiable advantage to software portability. Yes, it took some effort to port the system kernel from the Harris Datacraft to the PDP-11, and later to the VAX, but once this had been done, all application software could be used instantly. This was a real advantage as, at around this time, the late eighties, many institutions upgraded from PDP-11s to VAXes. Software portability all of a sudden became a way to protect one's software investment.

Not surprisingly this had been recognized by others, too. The most thorough of the efforts, which started then, was the Astronomical Image Processing System (AIPS[2]). Building on the experiences of the IPS, Don Wells and Eric Greisen at NOAO built this system for the radio astronomy community. It is my feeling that radio astronomers were more appreciative of new software technology, probably owing to the fact that they had started to use computers much earlier than the optical astronomers. Large radio telescopes typically had an altazimuth mount, requiring more than a clockwork mechanism to track the objects. And radio data were not readily visible, they needed to be rendered in a variety of ways in order to exploit and to interpret them. So radio astronomers had turned to computers earlier than the optical astronomers did.

This also happened in Europe, where Ron Allen and his group had built the Groningen Image Processing System (GIPSY).

7. Other Systems

This is not to say that nothing had been done in optical astronomy before the TV System came along.

The earliest "system" was probably VICAR, developed at JPL in support of NASA planetary missions. This was essentially a command interpreter, which turned human-readable and astronomically meaningful commands into IBM JCL. Having been funded by NASA, and being available, the system was also used to support the International Ultraviolet Explorer (IUE), which was launched in 1978. An interactive version was created which ran on PDP-11s rather than on an IBM 360, and the calibration pipeline of the IUE was implemented. However, the system was not transportable to the user community, and other software was used for the interactive processing of the IUE data.

The most successful one of these tools was the Interactive Data Language (IDL), the brainchild of Dave Stern, a clever and elegant system even on PDP-11s, which allowed the quick and easy manipulation of the spectra in a line-by-line command mode, as well as the generation of programs, using the same syntax, which could then be precompiled and executed much faster. This made developing and testing programs very easy, almost intu-

[2]See the chapter by E. Greisen in this volume. (Ed.)

Figure 6. Image processing work station in the computer room of the La Serena office of CTIO. The station was built in-house, using commercially available components (ca 1976). (Photograph by the author)

itive. A large body of application software was generated around IUE and was later used successfully for the first generation of spectrographs on HST.

IDL became a commercial product when Dave Stern founded Research Systems Incorporated (RSI). This was not popular with astronomers, who for inexplicable reasons abhor commercial software. A simple calculation shows that IDL was a good investment: the annual license fee was a fraction of the monthly salary of a software engineer. And, for this amount of money, one not only got an excellent software system, but also instant access to a large library of application software, most of it for spectral processing, worth tens of programmer-years. Incredibly, there were astronomers who were even opposed to using such software: "How you trust somebody else's flux integration?" – one step up from "How can you trust a number which you have not calculated yourself", but only a small step.

IDL was later ported to other and more powerful machines, allowing full processing of multidimensional images. RSI has since been sold and the new owners are less enthusiastic about astronomy, preferring instead to cater to the medical imaging market, where there is more money.

In Europe most of the interactive work with IUE data was done with

Image Handling and Processing (IHAP[3]), the first data analysis system produced by ESO. Most of it was written in HP 2114 assembly language, so it was not portable. However, it had become a *de facto* standard because it was used by all visiting astronomers at ESO telescopes and for their post-observation data analysis. So very pragmatically, HP machines were bought and IHAP was used for IUE.

The IUE satellite was operated by NASA, ESA and the British Science Engineering and Research Council (SERC – now absorbed into PPARC, Particle Physics and Astronomy Research Council). There were two almost identical ground stations at Goddard Space Flight Center, Maryland, and in Villafranca del Castillo near Madrid, Spain. There were regular three-agency meetings to coordinate. Yet the three agencies essentially ignored each other's analysis software development efforts. This sounds incredible today, but at a time when "downloading" meant receiving a 1600 bpi magtape in the mail, along with cryptic installation procedures, one pragmatically went with what was at hand and what could be supported locally.

8. FITS

There was one result which came out, at least in part, of the IUE experience, which worked to the benefit of all of astronomy, and which in many respects paved the way towards world-wide software sharing by demonstrating that benefits in terms of saving time and money could be derived from a little bit of coordination.

IUE data were delivered to the community on industry standard magtapes, plus on strip chart recordings and on large-format photographic films. Many astronomers worked off the strip charts, using pencil and ruler, not only because they were used to it from their previous work with ground-based hard-wired spectrum scanners, but also because quite often they could not read the magtapes. They either did not have tape drives, then a considerable investment for a small computer installation, and even if they did, more often than not they could not decipher the format of the data sets. Not that is was all that complex, but it was a mixture of ASCII, binary integer (for the raw data) and floating point (for the calibrated data) numbers, arranged according to the definition of the person who had created the writing software, and it was different for different modes of the instrument.

There had been a need for a standard image format to transport images even before IUE. In the mid-seventies it became painfully obvious that radio astronomers and optical astronomers could not share their data. Before that time this was not so important, astronomy had been divided

[3]See the chapter by P. Grosbøl and P. Biereichel in this volume. (Ed.)

along wavelength regions, and optical astronomers were using photographic plates and not magtapes. Only after microdensitometers and panoramic detectors became popular in optical astronomy did the data storage and data exchange issue become important. In the US, the National Science Foundation, who funded both optical and radio astronomy observatories, took the initiative and encouraged the astronomers to get their act together.

The very successful IUE satellite with its growing user community finally made the need obvious to everybody. Several people took the lead, most notably Don Wells and Preben Grosbol, and designed an image format for transporting images on magtape, the Flexible Image Transport Format (FITS[4]).

FITS has been a major achievement in astronomy. The secret of the success was probably that the designers started small, not overwhelming the user community and their capability to generate software. Changes and upgrades were small and incremental, backward compatibility was retained, and for every new version there was actual software delivered with it. Working software, which could be adapted to everybody's machine.

We have come to rely heavily on FITS. It saw the heaviest use in the mid-nineties, when the magtape-in-the-mail was the highest bandwidth computer-to-computer link. But even today the different archives produce FITS compatible data, albeit on different distribution media.

This high degree of success became obvious much later, when, during a recent internal ESA coordination meeting the topic of archive accessibility came up. The astronomers had not flagged it as an issue, in contrast to the Earth observing and multispectral analysis people. It turned out that, in remote sensing, different mission archives generally cannot read each other's data. That this had been relatively effortlessly possible in astronomy for more than 20 years came as a big surprise to that community.

9. The Faint Object Camera

In 1976 the European Space Agency (ESA) joined the Space Telescope (ST) Project. One of the European contributions was the Faint Object Camera (FOC). A Science Team was established and I joined as an expert for image data analysis.

During the discussions in the Science Team, it became obvious that ESA did not have data analysis facilities adequate to analyze the data that we expected to obtain with the Faint Object Camera. Realizing this, Duccio Macchetto, then the ESA ST Project Scientist, initiated a software production process by forming a team, which developed the specifications to go into an Invitation to Tender for industry. A Dutch software house

[4]See the chapter by E. Greisen in this volume. (Ed.)

won the contract and proceeded to build the Faint Object Camera Image Processing System (FIPS) with the FOC Software Team acting as reviewers and advisors.

Did a new system have to be produced? Yes and no. Of course there was the TV System, which I could have supplied, and which could have been upgraded. There also was GIPSY, and Ron Allen, also a member of the FOC Software Team, could have supplied it and it could have been upgraded. Of course the disadvantage of both systems was that they were community-supplied, in other words the software engineering was not quite as strict as for an industry product, and that the individual application programs relied heavily on the availability of the original author for their continued maintenance and upgrade. Right or wrong, the Agency felt that this was not a demonstrably reliable solution.

It also became obvious that, in addition to building spacecraft and instruments, one of the main functions of the various space agencies was to generate work for aerospace industry.

However, the FOC Software Team succeeded in creating, in Europe, a community-wide awareness of the growing importance of software in astronomy. Two national networks were started, partly as a result of the work of the Team, Starlink in the UK and Astronet in Italy. Both networks were extremely successful in supporting their user community. They have been alive as networks until well into the nineties, when eventually they were absorbed into the Internet.

It also encouraged ESO to launch the Munich Image Data Analysis System (MIDAS[5]), the long overdue successor to the IHAP system.

FIPS itself did not see much use. Following negotiations with NASA and with the European Southern Observatory, ESA contingents were established at the STScI and at ESO, where they used the respective local data analysis systems. These quickly surpassed FIPS in terms of functionality and level of support. Still, had things developed differently, ESA would have had a data analysis system ready to go, a prudent policy given the considerable investment for the FOC.

10. STScI

In 1976 the decision was made to build and launch the Space Telescope, later named the Hubble Space Telescope (HST). Following a recommendation by the US National Academy of Sciences the science operations were to be entrusted to a scientific institute, to be founded for the purpose. So NASA issued an announcement of opportunity, directed at large institu-

[5]See the chapter by K. Banse in this volume. (Ed.)

Figure 7. Barry Lasker (1939-1999) at the blackboard in his office at the Space Telescope Science Institute, then (early 1981) located in three rooms in Rowland Hall at the Homewood Campus of the Johns Hopkins University in Baltimore, MD. He is working out the strategy for the production of the Guide Star Catalogue, which was going to yield an unprecedented 20 million objects. Starting in 1995, Barry and his team re-did the Catalogue, compiling positions and photometric data for more than 435 million objects. (Photograph by Marjorie Felser. Courtesy STScI)

tions and consortia, for the establishment of what later was going to be known as the Space Telescope Science Institute (STScI).

AURA, the Association of Universities for Research in Astronomy, already operated Kitt Peak National Observatory (later to turn into the National Optical Astronomy Observatory – NOAO) and Cerro Tololo Interamerican Observatory (CTIO). So it was natural for AURA to decide to compete for the STScI contract. I was working at CTIO with Art Code at the time. Art was then the Chairman of the AURA Board and, together with John Teem, President of AURA, he was leading the proposal writing effort. I was invited to join and spent several cold weeks in Madison, Wisconsin, in January 1980, contributing to the section on data management. The driving force behind the effort was Barry Lasker, who was relentlessly pushing, and it was in no small measure because of his energy and dedication that AURA won the contract in early 1981.

Art Code, Barry Lasker and I went to work in Baltimore, Maryland, in

March 1981. We were supported by a small number of technical people and administrators from Computer Sciences Corporation, our co-proposer, and from Johns Hopkins University, our host organization.

Three of the science instruments of the telescope produced images or spectra: the Faint Object Spectrograph, the Faint Object Camera, and the Wide Field/Planetary Camera. The latter had four CCDs of 800 $\times$ 800 pixels each, enormous by 1981 standards. Expected data rate from the telescope was a whopping two gigabytes per day. Each and every day.

Under the contract, STScI was expected to produce the interactive Science Data Analysis Software (SDAS). Following standard procedures, NASA had convened an advisory team, the Space Telescope Data and Operations Team (DOT), headed by Ed Groth. The Team had done a large amount of preparatory work. It had, for instance developed the concept of standard interface routines, which would make it possible to easily absorb user-supplied software into a future HST data analysis system.

This concept would much later be re-born under the name of "ST Interfaces". However, owing to its early start and to the fact that everybody had a PDP-11, the recommendations of the DOT were biased towards it. We at STScI – and by mid-1981 this included Riccardo Giacconi and the team who had previously worked with him on the Einstein satellite – felt that a new approach was required. We organized a Data Analysis Workshop, which was attended by data analysis experts, by members of the teams who were building the Science Instruments, and by representatives of the space agencies. Based on the concepts, which were developed during the workshop, we then generated the requirements for the analysis software. These requirements, although rather modest when compared with the capabilities of even the image processing systems in use at that time, still called for a system, which was much larger than NASA had estimated. Getting the requirements approved was a major problem.

Why did we need "new" application software in the first place? Well, just like ESA in the case of the Faint Object Camera, NASA felt that community-supplied software was not good enough for the purpose. They were also concerned that the various contractors would charge large sums of money if they were forced by the Agency to accommodate outside software.

Data downlink from the telescope was done through channels, which were routinely being used for other NASA spacecraft, so getting the data to us was no problem. What NASA was – correctly so – concerned about was the question how to receive, calibrate, analyze, and archive the data, plus deliver them to the PIs on a time scale of much less than a week. Right or wrong, NASA did not trust the astronomers to be able to do this production pipeline in a demonstrably reliable manner. Instead the contract for what was termed the Science Operations Ground System (SOGS) was given to industry.

This caused a severe problem for the science data analysis software. Our software was intended to run on the SOGS, which meant that we were not supposed to write any code before the contractor was selected, had the design submitted and approved, selected the hardware, selected the host operating system, command language, etc.

Arguing that the astronomers we were hiring needed an interim computing facility we convinced NASA to give us a PDP-11, which was left over from another project, and we bought a 300 Mbyte disk drive the size of a washing machine (which was, incidentally, considered exorbitant, until we reminded everybody that this was less than 30 calibrated frames from the HST Wide Field/Planetary Camera, WFPC), installed the TV System and several other packages which our growing staff brought along, and started limping along.

In an effort to shorten the delay caused by the selection of the SOGS contractor we started to develop "generic" code for SDAS, which would plug into any operating system – provided it accepted Fortran, then a non-issue.

But we ran into an interesting problem. One of the requirements we levied upon ourselves was that the software be portable, or at least transportable, *i.e.* that it could be moved to other computers with little effort. After all, we knew that the community would want the software as soon as HST data were available. And even we ourselves would have the advantage of being to a high-degree independent of future platforms and operating systems.

This was however met with non-understanding during the various project reviews. It was pointed out that making the software transportable was going to drive the price way up, unnecessarily so. To calibrate this statement, one has to understand that driving up requirements, and therefore the contract value, is a common exercise for contractors, known as gold-plating. And the people who reviewed us were used to dealing with industry contractors. In fact, industry contractors are usually not keen on writing portable software – they want to be given another contract when the time comes to upgrade the software to the next version of the operating system. It took lengthy negotiations to arrive at a compromise which we called "limited portability", and which allowed us to start developing software for a data analysis system of which not even the hardware was known at this point.

Eventually the contractor was selected, hardware was specified (VAXes, not surprisingly) and delivered. It quickly turned out that we could not use the contractor-supplied data analysis host system because it was totally inadequate. It was obvious that it had been designed for a somewhat different user community. For instance, their "`TARGET`" command had the arguments:

latitude, longitude, and elevation above sea level. Lengthy negotiations ensued on what to do about this. As a stopgap measure we imported AIPS and ran our software from it. The concept of portability, albeit "limited" had paid off.

By the way, this is why there is an "epar" tool, so useful especially to new users of IRAF: Because we did not know on which host system our software would eventually run, we developed a host-system-independent way to communicate with the application tasks. If necessary we could, and we did, run SDAS straight from VMS.

11. IRAF, MIDAS, and the ST-ECF

When the SOGS-supplied data analysis host system turned out to be inadequate, other systems were investigated as possible candidates. Among them were AIPS, IDL, and the NASA developed Transportable Applications Executive (TAE). The system which was finally chosen was IRAF, the Image Reduction and Analysis Facility[6].

The advantage of IRAF was that it was being developed at NOAO, an AURA institute just like STScI. Not only was it felt that this would provide the possibility to influence the development of the system such that its use for HST data analysis could be maximized, it was also evident that the large community of American observational astronomers were going to be exposed to IRAF during observing runs at the National Observatories in Arizona and in Chile.

IRAF, which normally used its own precompiler for even better portability, was modified to accept Fortran-written software, and the SDAS software was salvaged by collecting it into an IRAF package.

By 1984 Piero Benvenuti and I had started up the Space Telescope European Coordinating Facility (ST-ECF). ESA has a 15% investment in the HST, and the ST-ECF was formed to support the European astronomers in the use of the Hubble Space Telescope. Based on the consideration that the European users of the HST would be the same community, which was using the large ground-based observing facilities of ESO, the ST-ECF was located at the headquarters of ESO in Garching near Munich in Germany.

ESO had, over the years, developed their own image processing system, the Munich Image Data Analysis System (MIDAS). By the mid-eighties the system was large and powerful, it was in use at many different institutions world-wide, and ESO was spending considerable efforts in supporting it.

[6]See the chapter by G.H. Jacoby and D. Tody in the earlier volume published in the same series (*Information Handling in Astronomy*, Ed. A. Heck).

In particular the system had reduction packages for data taken with the telescopes and instruments at La Silla, the ESO observatory in Chile.

Through accretion, or through the power of large organizations, the image analysis world had by 1985 split into essentially four large groups: IRAF at STScI and NOAO, AIPS for radio astronomy, MIDAS in continental Europe, and STARLINK in the UK. Coordination among these software development efforts was considered necessary.

One area in which a special need existed was the area of image display. These were the days when image display hardware was expensive, especially the large-format color displays. The devices were one-of-a-kind, making software exchange essentially impossible. The idea was born to develop a set of common low- and mid-level routines, which operated the hardware of the different devices, making it possible to share each other's high-level software. Unfortunately the effort did not succeed. In the first place we made the mistake of aiming too high: we were trying to define everything for everybody into the infinite future. If the FITS committee had tried to do this, nobody would have accepted it. Well, nobody really accepted the Image Display Interfaces either, to my knowledge there was only one implementation, in MIDAS. Then SAOimage came along, a freestanding package using the emerging X-windows, which could easily be connected to other software. And eventually the rapidly growing market for image displays, driven by video games and multimedia applications, made the need for Image Display Interfaces obsolete.

This was a difficult time for the ST-ECF. On the one hand, most of the HST-related software was generated in IRAF and, on the other hand, the European astronomers were used to MIDAS. Additionally, there was a large IDL community on both sides of the Atlantic. Several different options were investigated, but it was impossible to reconcile the different communities. In the end, we decided to be pragmatic, to support all three systems, and to contribute software to where it was most appropriate on a case-by-case basis. This did not please the purists, but this was all we could do with our limited resources

Starting in 1985 we organized annual Data Analysis Workshops in order to improve the general level of data analysis and in order to obtain feedback on required analysis tools from our user community. As this was the time when, unlike today, not all users had the computer capacity to analyze image data in their offices, the time around the Data Analysis Workshops was quite often used by the participants to reduce their ESO data on our data analysis facilities, thus turning them into true workshops.

Because of the importance of the HST to optical astronomy and the widespread usage of HST data, both through observations and through the use of data in the HST Science Archives, and, not least, through the possibility and ease of software downloading through the Internet, IRAF has become the de-facto standard for data analysis in optical astronomy.

12. Spherical Aberration

Most of the application software in the different image processing systems in the seventies and eighties were relatively straightforward, doing I/O, image manipulation and arithmetic, and image display. Sure, there were occasional sophisticated analysis algorithms for doing things like object extraction and photometry, but these were not the main components of the system. The main reasons for having a "system" were: independence of the vendor-supplied operating system, being able to use astronomically meaningful commands, provide input in astronomically meaningful units, meaningful error messages, the availability of efficient I/O routines which were optimized for pixels, a consistent "look-and-feel" across all this wide range of software, and, as already elaborated, the ability to share software across operating systems.

This changed dramatically in mid-1990, when it was discovered that the main mirror of the HST suffered from spherical aberration. It was obvious that the problem was serious and would eventually have to be corrected through a hardware fix. In the meantime, however, efforts had to be made to improve the quality of the HST images by ground processing.

Contrary to a widely-held belief at the time, the HST images had not lost resolution. The central peak of the point-spread function was as narrow as had been expected. It was just much lower than expected, most of the energy having been spread into the wings of the PSF. This amounted to a loss of contrast and thus to a loss of limiting magnitude. But it meant that the full resolution of the images could be recovered by deconvolving the data with the point spread function.

This had been done even much earlier with considerable success. At the occasion of one of my first visits to Kitt Peak in 1970, Jim Brault showed the Fourier filtering and deconvolution of solar spectra. We even briefly considered doing something similar with the Lowell Observatory Data System, but we quickly ran into the hard limits of our PDP-11. Also, solar spectra have the considerable advantage that there is no shortage of signal, *i.e.* signal-to-noise is as good as one desires.

Given the importance and the visibility of the HST, considerable efforts were made to develop good image deconvolution routines. Within a short time a veritable deconvolution industry started up. And the emphasis of the image analysis software development shifted from the system to the applications.

Of course the spherical aberration problem was a serious setback for space astronomy and, indeed, it caused enormous problems for NASA, coming right at a time when the Shuttle program had barely recovered after the Challenger accident. However, at the ECF we decided to consider it a challenge rather than a defeat. Since it was obvious that the serious processing required to solve the problem needed large computing resources, we immediately upgraded our computer systems. We also hired Leon Lucy, who had years before developed what is now known as the Lucy-Richardson algorithm for the non-linear restoration of image data. Together with Hans Martin Adorf, Richard Hook and Fionn Murtagh, he succeeded in generating and making available to the astronomical community a wide range of image restoration tools, much appreciated at the time.

Although the restoration requirements have gone away after the correction of the spherical aberration through the Corrective Optics Space Telescope Axial Replacement (COSTAR), the software still serves us well for other purposes. Fusing data with different point spread functions, for instance ground-based and space data, is possible, as is the selective removal of different sources of contamination, for example sloping backgrounds or cosmic rays. Finally, the resampling used in methods like drizzling and pixel subsampling has its roots in the restoration software.

13. Software Coordination in the Nineties

It is very difficult to compare computing over a 30-year period. During the last 10 years, typical astronomy department computing went from hundreds of users sharing a single computer to each one of those users having a computer on their desk which is a thousand times more powerful than the old single shared one, and comes at $1/1000^{th}$ of the cost. During this time the computer has evolved from being an accessory to being the main tool of the astronomer, much more so than the telescope. It is now possible to do meaningful astronomy without a telescope, but it is not possible to do meaningful astronomy without a computer. The Astrophysical Virtual Observatory in Europe and the National Virtual Observatory in the US are the natural consequences of this development.

However, some basic considerations are still valid, such as coordination of the software development effort to maximize the use of the available resources.

We like to think that the reason for the high level of coordination among the astronomers is that we are somehow more able to think in global terms. Of course there is also the fact that there is no commercial interest and no political or strategic motivation in astronomical research. Be that as it may, astronomers have pioneered the use of international computer networks for science. As early as 1985 we had computer-to-computer communication working between the STScI and ESO/ST-ECF. I already mentioned STARLINK as a national network. The first European International Computer Network was SPAN, the Space Physics Analysis Network. And by the beginning of the nineties world-wide e-mail communication among astronomers was an established fact.

So it was not by chance that astronomers embraced the World-Wide Web as soon as it became available for practical use in 1993. The Web servers at NRAO, CADC, and ST-ECF were among the first 200 registered web servers world-wide! And by the beginning of 1994, astronomers were the largest single user group on the Web, a fact that was noted with some surprise by the Wall Street Journal.

Astronomical Data Analysis and Software Systems (ADASS) Conferences have been held with considerable success since 1991, taking the ECF-organized Data Analysis Workshop to a world-wide forum. It is probably correct to say that astronomy is the science with the highest awareness throughout the community of the software tools available, and with the highest level of software sharing.

In many ways, the software sharing and software coordination effort is less obvious today than it was a decade ago: it happens, it happens without much ado, and on the basis of well-established cooperations. For instance, the fact that the software for the analysis of data generated by the spectral modes of the NICMOS and ACS instruments on the HST was developed at the ST-ECF only becomes visible to the users if they need software support and their request is forwarded to us by the STScI. Obviously all data analysis software is available in both places, and in others, such as the CADC. And everybody runs mirror sites as a service to their respective communities. Astronomers have finally overcome the "not invented here" and "gotta do it myself" attitude. Well, most of them, anyway.

14. Lessons Learned

What could have, should we have been done differently? Not much, I think. Sure, NASA and ESA could have saved a considerable amount of money if they had not taken the detour through industry contracts for the HST software development. But that would have taken clairvoyance on the part of the managers in charge, or at least a high degree of trust in the abilities of the astronomers to produce the required software on time and within budget. What was done was the prudent thing to do under the circumstances.

Of course the astronomical software developers improved too: the level of software engineering is without doubt much higher than it was in the eighties. Quality assurance, configuration control, testing and documentation are issues, which we all accept as important for the success of our software. We do have the advantage that we work close to our customers, and sometimes we ourselves are those customers. So we can afford to take shortcuts, push some of the testing out to the users, for instance, and thus make the software development process more efficient.

Still, the software development process is lengthy and costly. We need to give our users high-level tools, which allow them to tailor their data analysis procedures quickly and easily. By the same token, we must not allow them to influence procurement decisions by the need to run obsolete code. Jump the gun, re-code, or reverse-engineer, document the software and insert it into IRAF. It will then live as long as IRAF is supported.

Commercial software is being used with less hesitation in astronomy these days. I know that not everybody will agree with me on this, and there are certainly negative examples, but we have to recognize the fact that labor is the cost driver. Any software development of significance will take several person-months, *i.e.* an investment of thousands of Dollars. Most license fees for commercial packages cost less than that, and they come with service support, generally decent documentation, and a library of applications software. Sure, there will always be a need for special-purpose software. But the development has to be seen in analogy to the history of electronics engineering in astronomy.

Until well into the seventies astronomers used to build their own DC amplifiers. In infrared astronomy this trend continued into the eighties. Initially this had to be done because there were no commercial products available which could do the job. But today nobody should build their own CCD camera – industry delivers products which in almost all aspects are superior, and definitely cheaper, than what an astronomer can do in an observatory electronics laboratory. And, just like there is still a need for special-purpose software, there is the occasional need for special-purpose hardware, such as when building instruments for use at very large, or very high-flying telescopes.

A very touchy issue is the proper use of new technology. We know that the technological evolution proceeds very rapidly. Astronomers, being unusually clever, follow this evolution and quite often latch on to new technology earlier than others. However this sometimes works to our disadvantage because new technology tends to be very expensive, and sometimes it becomes obsolete very quickly. An example is the use of optical bulk storage media for the science data archives of the HST. Initially planned as a tape archive, we converted to optical media as soon as they became available. We fought with hardware installation and maintenance problems, even wrote our own driver software. Soon after we had the archive working, and before the HST was actually launched, a new generation of optical disks came along, the old hardware became obsolete and unsupportable, and we had to convert the archive.

And again several years later. We realized with considerable surprise that the useful lifetime of the archive media was a fraction of the lifetime of the observing facility. We finally broke that vicious circle in 1996 by falling back to what many then considered obsolete technology, which, however, was firmly embedded in a world wide market, so it was cheap and there was guarantee that it would be around for at least another decade: compact disks. CDs were in everybody's living room, and while the technology might not have been cutting edge any longer, it was certain that nobody would accept having to throw out their CD collection. And even though the capacity of a CD-ROM was only about a tenth of that of a large optical disk, the price of the raw medium was only about one hundredth. Plus the price of the readers and burners, and of the robotic devices was eminently affordable. We also saved capacity by junking the calibrated data and installing extra computer capacity to do on-the-fly recalibration. In the meantime we have, of course, upgraded to DVDs, but only after they had firmly established themselves in the market. The lesson there is: unless you have very special needs do not push technology, instead let technology push you. Something to keep in mind when looking at the oncoming GRID bandwagon.

The final lesson is that we have demonstrated that the astronomers can get together and collaborate on a global scale, not only in their quest to understand the universe, but also in the much more mundane, but quite often equally demanding area of building the tools which make the noble academic goal possible.

Acknowledgements

I would like to thank Karl Rakos, Peter Boyce, John Wood and Skip Schaller for their support of my work, for helpful comments on this paper, and for remembering details, which I had forgotten. I would also like to honor the memory of the late Barry Lasker, who during the almost 30 years that we knew each other always supported, often contributed to, and sometimes challenged my efforts through ideas, advice, and constructive criticism.

IHAP:

IMAGE HANDLING AND PROCESSING SYSTEM

P. GROSBØL AND P. BIEREICHEL
European Southern Observatory
Karl-Schwarzschild-Straße 2
D-85748 Garching, Germany
pgrosbol@eso.org

Abstract. The IHAP system was conceived by Frank Middelburg around 1973 after he moved to the ESO Headquarters in Geneva from Chile where he had become a computer expert making several instrument control systems. It was implemented on HP 21MX series computers for which it was highly optimized. IHAP used its own file system for image data and mainly assembler code for low-level routines while applications were written in FORTRAN. This gave it a very high performance considering computers at the time but made it difficult to port. It contained most of the features of a modern data processing system such as interactive and batch modes of operation, world coordinates for images, a table system and a device independent display manager. The system was employed extensively at La Silla for quick-look and on-line reduction. For off-line reduction, it offered a full set of reduction procedures for spectra and images. It was also exported to 15 major institutes in Europe. With the availability of cheap work-stations and powerful 32-bit mini-computers in the mid 1980's, it started to yield to MIDAS which had taken over many features and applications from IHAP.

1. Introduction

From the creation of ESO in 1962, it was clear that computer-based control of telescopes would be important. Concrete suggestions for computer automatization of the ESO 3.6m telescope, planned to be erected at La Silla, were made in 1968 and included control of telescope, dome and instruments.

Digital computers were first introduced at ESO in 1970 when it was decided to control the one dimensional GRANT measurement machine, then

A. Heck (ed.), *Information Handling in Astronomy – Historical Vistas*, 61-70.

situated at the ESO office in Santiago, Chile, to gain experience with digital automatization of instruments. At the same time Frank Middelburg, who joined ESO in 1967 and worked as night assistant and observer at La Silla, had become interested in the new possibilities offered by computers and started learning system design and programming of the HP 2100 mini-computers chosen by ESO for telescope and instrument control. He designed and implemented the computer-based control system of the GRANT machine in collaboration with several ESO astronomers (*e.g.* J. Rickard and A. Ardeberg). The system formed a basis for many instrument control systems at La Silla.

2. Start of IHAP

A few years later, Frank moved to Europe to join the Telescope Project division at the ESO Headquarters which had been established on the CERN premises in Geneva. There, a two-dimensional S-3000 measuring machine was installed and used to digitize astronomical images from photographic plates obtained at ESO including the Schmidt telescope with its large format plates. He early realized the importance of not only using computers for control of telescopes and instruments but also for processing the data acquired. In 1973-1974, he started unofficially to design a system for *Image Handling and Processing* (IHAP) to reduce and analyze astronomical spectra and images mainly obtained by digitizing plates on the ESO measurement machines. By the end of 1974, technology had advanced enough to make it feasible for ESO to consider purchasing an image display device which could show digital images. Eventually, it was decided to acquire an IMLAC graphic display with a 21-inch screen which could show 20K pixels with 16 grey levels. It was delivered to ESO in early 1976 and interfaced to the IHAP system which by this time had become an official project. A few years later, as technology advanced, the IMLAC display was replaced with RAMTEK systems which had 512×512 pixel resolution with 10 bit color levels controlled by a 12-bit by 2048 words color look-up table for pseudo-color representation in addition to overlay and text channels.

Digital detectors such as the Image Dissector Scanner (IDS) on the Boller & Chivens spectrograph at the 3.6m became available in 1978 and made it essential to perform simple reductions and display results directly during the observations. The IHAP system was well suited for this purpose and became the default quasi real-time, quick-look data reduction system for La Silla. In fact, IHAP included modules to interact directly with instruments and perform data acquisition.

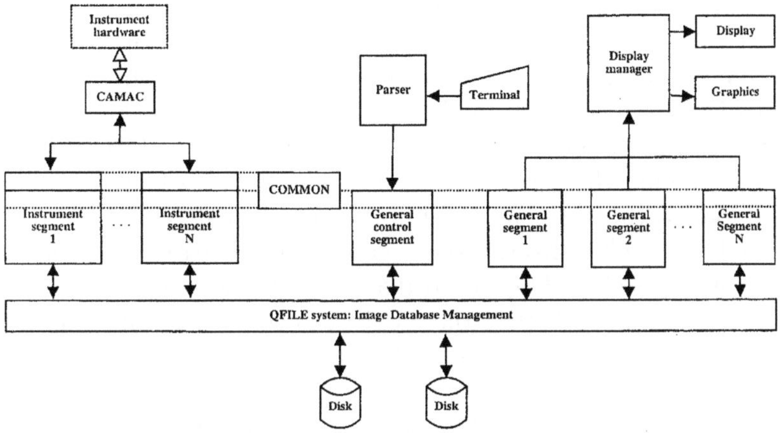

Figure 1. Schematic design of the IHAP system

3. System Configuration and Design

The IHAP system was based on HP 21MX computers running the real-time, multi-tasking operating system RTE and typically equipped with up to 256kb memory, floating point unit and 100Mb disk storage of which about 25Mb were used for the RTE system, programs and general user data. Besides alpha-numeric terminals for user interaction, the main devices were HP2648 graphic terminals for plotting, HP7221A pen plotters, RAMTEK image displays, magnetic tape drives and line printers. More than one data reduction session could be executed at a time due to the multi-tasking capability of the operating system.

A major challenge was to create an interactive image processing system with a reasonable response considering the speed of mini-computers in the later part of 1970's. Although digital detectors were still relatively small, the system was conceived to handle images from digitized plates with sizes up to 2048×2048 pixel which placed high demands on both CPU and I/O channels. The RTE operating system executed tasks in 58kb segments which also meant that even relative small image could not be resident in memory but had to be accessed sequentially. These limitations demanded that all available resources of the system were utilized to the extent that several low-level parts were coded in assembler language and a special I/O subsystem (*i.e.* the `QFILE` manager) was designed for optimal transfer of image data to and from the disk. The FORTRAN language was used for high-level routines and user applications.

The basic components of the IHAP system is shown in Fig. 1 as a block diagram. When the system started the *General Control* segment took control and waited for user input after it had performed all necessary initiations. All user input was handled by a *parser* which converted the user provided command parameters into internal representation. Once the *General Control* module had obtained the user input from the parser, it placed all information in a general COMMON area (in the FORTRAN sense), found the segment which contained the code to be executed for the command and loaded the segment. All information transfered between segment were located in the COMMON area except for large arrays for which system disk tracks were used. A *Display Manager* with a generic interface handled all graphics and image display output from user applications and thereby isolated them from the detailed control sequences needed to operate the specific devices.

Instrument control was performed by a special set of segments which in addition to the general COMMON also communicated through a separate instrument COMMON for security reasons. These modules interfaced to the instrument hardware through a CAMAC interface. The *Instrument* segments could both send commands and receive data from the instruments which then were stored on the IHAP disk areas.

All image data transfered between segments and the disk went through the image database manager named `QFILE` which optimized I/O for random access storage devices. It supported two directories of files where one was loaded into core for fast indexing of the data, while the other was located on disk and contained the description of the images. The `QFILE` manager buffered data so that data falling between sector boundaries on the disk was kept during the I/O operation in such a way that only a single disk seek was required for a given transfer. Since such disk seek operations were relative slow, this feature provided a substantial speed improvement. Further, when more than one disk drive was available the manager ensured that input and output images were placed on different drives to avoid competing I/O operations. Data were stored sequentially on raw disk tracks which also improved access speed as files by definition could not be fragmented. To recover disk space from deleted files, the user had to issue an explicit `PACK` command which would reclaim space of purged files by physically copying files.

4. System Features

The system handled 1D spectra and 2D images in either 16-bit signed integer or 32-bit floating point format where the integer format was primarily used for raw data. The two formats were transparent to the user and integer files were automatically converted to floating point when floating point operations were performed on them. For efficiency, sets of 1D spectra could be stored in a 2D frame and accessed individually as so called scan-lines.

References to image files were made by numbers which the system allocated sequentially to each new file created by a command. Files could also be tagged with a name which however only was kept within a user session.

The concept of linear world coordinates for both spectra and image was an integral part of the system from the start. The reference to a pixel was standardly given in these world coordinates as a set of real values which then internally were transformed to pixel indices by dividing by a floating point step size and subtracting a starting coordinate. World coordinates were used whenever images were related to each other such as arithmetic operations between images. If the sampling grid of the images did not agree to a fraction of a pixel, an error would be issued.

Images were stored on disk in an internal format which contained a fixed sized header with data identifier, information on type and size of the data matrix, celestial reference coordinates, time of creation, some user defined data, display scaling and a limited area for variable length ASCII records for user comments. After the header data were placed in 8kb blocks which gave an efficient storage when they were written to magnetic tapes.

The user interaction with the system regarding the control language, batch facilities and standard operation mode is described in the following sections.

4.1. COMMAND LANGUAGE

User requests were given, as in the RTE operating system, by command lines consisting of comma separated strings in the form:

```
[LABEL:] COMMAND, P1, P2, ... , P12          comment
```

where `COMMAND` was a string with the name of the IHAP task to be executed and `Pn` were command specific parameters. An optional label could be placed in front of command name while anything appearing after 5 or more consecutive spaces was regarded as a comment. Besides numeric and string parameters, special meaning was associated to numbers prefixed by a specific character such as '`S`' for scan-lines, '`X`' or '`Y`' for world coordinates, '`%`' for percent, '`#`' for file numbers, '`G`' for globals (see below) and '`K`' for column keys in tables (see below). The system kept the last 48 commands which could be recalled and edited.

Whereas approx. 100 commands were available in 1977, already by 1980 this had almost doubled. Besides system and display commands, the astronomical interesting commands covered a wide range and could be divided into the following general groups:

Spectral reductions: Tasks for wavelength and flux calibrations of 1D spectral data. This included functions to perform non-linear rebinning and comparisons with standard observed with different sampling. Special functions corrected S-distortions in long slit spectra and extracted orders from echelle spectra.

Image Transforms: A wide variety of routines for both sampling and intensity transformations. Images could be aligned by a linear transformation whereas different sampling functions were available for spectra. Besides standard numeric filters, Fast Fourier Transforms (implemented by K. Banse) and Lucy deconvolution (written by P.Grosbøl) were offered.

Fitting: Procedures to fit different functions to data. For spectral data, multiple Gaussian fits to lines were available while the continuum could be approximated by either polynomials or spline functions. Several methods for estimating the background variations on images were implemented.

It is worth mentioning the `PFUNCTION` command which performed general image arithmetic on one or more images and used a polish notation. The world coordinates of the images involved in the operations were compared and the result contained only the overlapping area. In addition to standard arithmetic operations, trigonometric functions were included using radians. Since special floating point values (*e.g.* not-a-number and infinity) were not implemented by the HP hardware, floating point errors (*e.g.* division by zero) returned a zero value.

Most commands which generated result would both print them out on a reduction log and save the numeric values in special system variable called *globals*. Space for 1500 globals was allocated with each global having type and value. This made it possible to transfer results from one command to a following one which was essential for writing batch procedures (see below). Some of the globals were reserved for special purposes such as `G0..G9`: input parameters for batch procedures, `G10`: number of last file created, and `G12`: result of last `PFUNCTION` command.

4.2. BATCH PROCEDURES

An extremely useful feature of IHAP was its ability to execute sequences of commands stored in ASCII files on disk. This was done by a special command:

```
BATCH, filename, cartridge, P1, P2, ... , P10
```

where the `filename` defined the ACSII file on disk `cartridge` and the parameters `Pn` were made available for the batch procedure via the globals.

Besides the normal commands, batch procedures had access to a set of special flow control commands such as:

– BIF: performed a comparison between two parameters and transfered control to the command with the specified label depending on the result of the comparison.
– BDO: was equivalent to a FORTRAN DO loop and was terminated with a BEND command. The loops could be nested to 16 levels.
– BGOTO: provided an unconditional jump to the command with the label specified.

By using the globals, it was possible to construct a long sequence of operations which otherwise had to be executed by hand. Especially for reduction of standard data (*e.g.* from IDS or measuring machines) this greatly improved the efficiency for users.

As a simple example of a batch procedure, we show one which reads N images from a magnetic tape, subtracts a bias, divides with a flat-field and finally displays the result. The procedure called CCDRED would be executed by the command:

```
BATCH, CCDRED,, 30, #4, #5
```

where 30 files will be read and the bias and flat-field are stored in the files #4 and #5, respectively. The procedure could look as follows:

```
BCHECK, 1, 4, 4                   check input parameters
BIF, GO, LT, 1, STOP              go to STOP if N<1
BDO, , 1, GO                      loop from 1 to GO
 RFITS                            read FITS file from tape
 PFUNC, G10, G1, -, G2, /         compute (G10-G1)/G2
 KDISP, G10                       display new file
BEND                              terminate loop
STOP:BTERM                        terminate procedure
```

which illustrates the way one could write procedures for IHAP. Some commands (*e.g.* BIF) could be omitted but are included to illustrate the syntax. For an actual procedure, one would include additional checks and possibly purge redundant files and pack the disk to regain space.

4.3. USER SESSIONS

The cost of computers with sufficient memory and display systems was so high that only very few image processing stations were available. At ESO, 2-3 users could work simultaneously and only few other institutes had similar facilities. This meant that many ESO observers also went to the ESO Headquarters to use the IHAP system to reduce their data which further increased the pressure on the facility.

Typical users would be able to reserve a few hours on an IHAP station per day depending on time of day. Since each station only had 50-100Mb on-line disk storage, users normally had to read all their data in from magnetic tape when starting the session and save the results afterwards as the IHAP disk areas by default would be cleared in the start of each session.

5. Hey-days of IHAP

By the end of the 1970's, IHAP was the default data processing system for ESO data and used for all new instruments at La Silla for quick-look and on-line reductions. It contained a wide variety of tasks for reduction and analysis of astronomical data and was even used to reduce other types of data *e.g.* IUE spectra. Further, several external institutes were obtaining copies of IHAP and used it at their home site (*e.g.* Bochum in Germany).

For the analysis of large quantities of data, it became clear that one needed to store results in tables *e.g.* magnitudes of stars or properties of spectral lines. The issue at hand was to treat collections of records each with the same type of information and structure. As a result, the IHAP table file system was created by F. Middelburg in discussions with P. Grosbøl who was interested in using it for doing photometry of galaxies. A table consisted of a set of rows which could be addressed either by its absolute position or through a numeric identifier. Rows had an identical structure defined by the columns which could be accessed through the *key* number (*e.g.* `K21`). Tables contained only real values and were limited to 41 columns and 32k rows. Many commands could be applied to table data such as sorting rows after values in a specified column, fitting of functions to values in columns, performing arithmetic operations between columns and merging or copying data. Many commands would be able to save their results in tables by appending them *e.g.* the search command which identified objects on images would add each new object found to the table.

The number of applications increased steadily to almost 300 in the mid 1980's. The system was at that time distributed to 15 major institutes which used it for their internal reductions. By that time IHAP had become so big that, for security and configuration reasons, it was decided to separate the instrument control functions and the pure data processing parts.

6. Time of Transition

One great strength of IHAP was its very high performance considering the computers at the time. However, it also became a significant problem as it was achieved by use of very machine specific code (often in assembler) which was difficult to port to new and faster systems. It was ported to the HP A900 system which was rather similar to the 21MX series but major parts would have had to be rewritten for a migration to significantly different architectures.

Another problem was the somewhat complicated structure of application programs due to the small segment size. This made it very difficult for normal astronomers to write new applications programs for IHAP.

During preliminary discussions on the software systems for the ESO 3.5m NTT telescope, Frank proposed around 1983 to make a portable IHAP-II system. At that time, MIDAS[1] had already been developed to a useful level and it was decided to concentrate efforts on that system because it was based on a more general purpose design which was less machine specific.

After the untimely death of Frank Middleburg in November 1985, Peter Biereichel took over the maintenance of IHAP. He assisted users at the IHAP stations, re-wrote the IHAP User's Manual and provided for new IHAP interfaces to the ESO real-time data acquisition system (AsterX). However, only few new image processing features were added to IHAP due to the limitations of 16-bit computers to which IHAP was bound. It was clear that the next generation of image processing systems had to run on the new 32-bit architecture machines in order to cope with the increasing size of detectors.

La Silla continued using IHAP for many years after Frank passed away, since it was an extremely stable and fast system. New instruments (Boller & Chivens, CASPEC, IRSPEC, F/35) and telescopes (2.2m, NTT) were using IHAP for quick look image and spectra display during data acquisition. This feature increased enormously the observing efficiency. IHAP was involved in performing automatic observations, such as acquiring a guide star, centering a star into the spectrograph slit, telescope focusing, etc. The dream of an observer to prepare the observations before a night and just watch (or sleep) while they were running became reality at the 2.2m telescope one year after it went into operation.

7. Conclusions

The IHAP system was conceived and implemented almost single-handed by Frank Middelburg at a time when few were dealing with astronomical data processing and even fewer imagined that mini-computers were capable of doing it. His interest in interactive images processing in close relation with data acquisition significantly helped the ESO community to optimize their usage of the telescope facilities and better cope with the ever increasing amount of data.

[1]See the chapter by K. Banse in this volume.

The IHAP system contained most of the basic elements required for a modern image processing system such as interactive and batch modes of operation, world coordinates for images and treatment of collection of records in the form of a table system. It was written and optimized for the HP 21MX series of computers. This gave it a very high performance by using assembler code for low level routines and its own optimized file system for images (*i.e.* the `QFILE` manager) but on the other hand made it very difficult to port to other computer architectures.

The data processing experience and many explicit features of IHAP were used for MIDAS. Notably, the *Table File System* in MIDAS was heavily based on the IHAP experience. Also many applications were ported from IHAP to MIDAS. As such the IHAP system formed the basis for image processing at ESO and was essential for the observatory for both quick-look and off-line data reduction.

Acknowledgments

We would like to thank K. Banse, J. Breysacher, W. Nees, R. West and M. Ziebell for discussions on the early history of IHAP. Information in articles published in the ESO Messenger, the ESO Bulletin and the book on *ESO's early history* by A. Blaauw were also used to obtain the correct chronology.

FITS: A REMARKABLE ACHIEVEMENT IN INFORMATION EXCHANGE

E.W. GREISEN
National Radio Astronomy Observatory†
P.O. Box O
Socorro, NM 87801-0387, USA
egreisen@nrao.edu

Abstract. The FITS format is a remarkable achievement in information handling and sharing. Astronomy is alone among the sciences in having an international data interchange format that is used by virtually all scientists and institutions in the field. The technical and sociological reasons for this success are discussed and a few of the many remarkable scientific results made possible by this information handling are described.

1. Early History

Wells & Greisen (1979) wrote that "With the advent of the WSRT and the VLA in radio astronomy, the increased use of CCD arrays and other digital techniques in optical astronomy, and the development of satellites for astronomical observations at other frequencies, the number of images in digital form has increased enormously."

This sentence has become an immense understatement. The need for scientists to carry data between observatories and to their home institutions has kept pace with the progress in astronomical instrumentation. Research projects now normally involve multiple wavelengths and multiple instruments and remote observing has become far more common. In 1979, each institution typically had one or more software packages tailored to its instruments and computing facilities. Almost every institution had developed at least one unique data format, for both internal and external data repres-

†The National Radio Astronomy Observatory is a facility of the (US) National Science Foundation operated under cooperative agreement by Associated Universities, Inc.

A. Heck (ed.), Information Handling in Astronomy – Historical Vistas, 71-87.

entation, and a significant body of software based on the use of that format. If N institutions wished to exchange data, a total of $N(N-1)$ programs had to be written to perform the translations between these formats.

When one of the institutions changed its internal format (and that happened frequently), then the other institutions had to make corresponding changes in their translation programs (if they were able to determine what change was to be made). With the development of a usable interchange format of any sort, this situation is improved dramatically. Each institution needs to write only two programs, those which translate between the internal and the interchange format. And when an institution changes its internal format, it alone is responsible for making corresponding changes in its translation programs. Of course, if the internal format *is* the interchange format, then no translation programs are required.

By 1978, a number of us responsible for major data centers had become tired of writing and rewriting data translation programs and had taken tentative steps toward developing and using a more general format. As early as 1976, Ron Harten (Netherlands Foundation for Radio Astronomy) proposed the use of a transport format. Don Wells (Kitt Peak National Observatory) and Harten subsequently exchanged data in two prototype formats. Following discussions involving representatives of the National Radio Astronomy Observatory (NRAO), KPNO (now National Optical Astronomy Observatory), and the National Science Foundation, W. R. (Bob) Burns of the NRAO wrote a memo in February 1979 urging that active discussions begin on an interchange format. This resulted in a meeting held at the VLA site March 27 and 28, 1979 in which the FITS format was designed by Wells and Greisen with advice from several NRAO staff members at the VLA, particularly Barry Clark. The first magnetic tapes to use the new format were exchanged in April 1979.

The first FITS files were written by a PL/I program on an IBM 360 under OS/MFT (32-bit, twos-complement numbers and 8-bit EBCDIC characters) and were read by a Fortran program executing on a CDC 6400 under SCOPE (60-bit, ones-complement numbers and 6-bit "Display Code" characters). This was very near the worst possible combination of environments and yet the interchange worked as intended on the first try. The FITS paper was presented in Trieste by Wells and Greisen in June (1979). In October, Greisen and Harten got together in Holland and developed the "random groups" extension to FITS. That extension involved a more general view of what constituted an image and, with hindsight, should have caused us to change the basic FITS design. It was argued that too much time had already passed and so we chose to make the extension unpleasantly *ad hoc* in order to avoid making obsolete any files already written.

This was the first application of the guiding principle "once FITS, always FITS," a principle that has been one of the main reasons for the widespread acceptance of the format. The two papers were submitted to the *Astronomy & Astrophysics Supplement Series*, appearing in June 1981 (Wells *et al.* 1981, Greisen & Harten 1981). By that time, FITS had already become the de facto interchange format for astronomy. Recognizing this fact, the Chairman and Co-Chairman of Commission 5 of the IAU, Bernard Hauck and Gart Westerhout, asked this author to recommend a resolution for Commission 5 at the 1982 meeting in Patras, Greece. It was adopted (IAU, 1982) and a working group to develop further extensions to FITS was established under the leadership of Preben Grosbøl.

2. Basic FITS

The acronym FITS stands for Flexible Image Transport System. This name expresses its main goal – to be a *flexible* means by which *image* data (by now *information* would be a better term) may be *transported* between co-operating computer systems. However, that very flexibility required the development of a quite general way of thinking about data and about the means by which they may correctly be described.

One of the first key decisions in the development of FITS was the selection of the length of the logical record. The choice of 23040 bits (2880 bytes) may seem strange now, but this number is evenly divisible by both the byte and word lengths of all computers that have been sold on the commercial market. It is small enough to be handled by the computers common circa 1979, but large enough to be efficient in writing data to magnetic tape, one record per block. All information for a particular "image" is contained within one file, either on magnetic tape or (now) on disk.

In FITS, character information is represented in 8-bit ASCII form; the other character formats common in 1979 are not allowed. Binary data were initially represented as 8-bit unsigned integers, and 16- and 32-bit twos-complement integers. Since that time, an IEEE-specified floating-point format (specifying the meaning of the bits but not the byte order) has come into wide-spread use in computing, and is now allowed in FITS files. The decision to represent the data in a binary format was initially controversial. However, formatting numbers, which are in binary form within the computer, into ASCII on the transport medium is inaccurate, expensive in computer time, and uses at least three times as many bytes on the output medium. Even in the computers of 2002, formatted reads are surprisingly expensive while the volume of the data has increased enormously. The bytes within the integer and floating numbers are in the order of decreasing significance (so-called "big-endian"). The reverse order ("little-endian"), used internally by many computer architectures including Intel personal computers, is not allowed; the time required for byte swapping is, however, negligible on modern computers.

The information in a FITS file is contained within one or more "header-data units" or HDUs. These consist of one or more logical records giving header information in the form of 80-character ASCII card images, 36 per record. Each card image contains an 8-character keyword (in upper case), usually followed by an equals sign and a value. There are a few required keywords which must occur at the beginning of the first header record. These identify the file as a FITS file and describe the binary format and dimensions of the data portion of the HDU. The required keywords are followed, in any order, by optional keywords, some of which are described in the FITS papers, and are terminated by an END keyword. The last header record is padded with ASCII blanks to its full 2880 characters. FITS writers are allowed to make up any keywords they may require, which allows the format to grow and to adapt to unforeseen developments. Although this is essential to the format's flexibility, it causes a failure to communicate. Until the meaning of a new keyword is described widely, most reading programs will have no idea how to interpret it. However, the header is always human readable and so, with adequate comments and history cards, may often be understood in time.

The data portion of the HDU begins in the first byte of the first logical record following the header record containing the END keyword. The data are a fully packed byte stream broken into logical records with no padding, except that the last data record is padded with zeros to its 2880-byte length. The initial data form described by Wells *et al.* (1981) was an n-dimensional, regularly spaced array. The arrays were described by giving the number of axes and the number of points on each axis in the required keywords. Greisen & Harten (1981) extended this data model to groups of such arrays each preceded by a number of binary "random parameters" describing the array. An example of a random-groups format would be a set of small images surrounding a variety of celestial positions with the random parameters describing the location of those positions. Although the random groups form has been widely used for radio interferometric data, it has largely been replaced by the binary tables form to be described below.

3. Standard FITS extensions

Wells *et al.* (1981) added a great flexibility to the FITS format by specifying that any number of 2880-byte logical records may follow the defined HDU. As might be expected, this led to the development of a variety of extensions, known and accessible only to their inventors. In order to provide a more orderly method for defining conforming extensions to follow the basic HDU,

Grosbøl *et al.* (negotiated in 1984, published 1988) defined a small set of new keywords and a general data structure very much like the image and random-groups data structures to be used in future conforming extensions. They stated that "The most important rule for designing new extensions to FITS is that existing FITS tapes must remain valid." The extensions have the same HDU structure as the primary HDU. The header contains required keywords at the beginning to name the type of extension and to describe the binary format and dimensions of the data portion of the HDU. This enables reading programs to determine the type of extension and the number of binary data records that must be read or skipped after parsing only a few keywords. The structure defined by Grosbøl *et al.* (1988) allows for any number of conforming extensions to occur in the data file in any order. The association of the data in the extensions with the data of the primary HDU and each other is indicated by their presence within the single file (Grosbøl *et al.* 1988).

A number of standard extension types were developed in the next few years. The first, by Harten *et al.* (1988), was a companion paper to the general description. It defined a means by which tabular data such as catalogs could be transmitted in a FITS extension in a self-documenting data structure using a fully printable ASCII form. Despite its inefficiency, ASCII tables have been very successful in the exchange of simple catalogs, and provided a way to wrap old representations in a portable framework. The immediate human readability remains an advantage. The second conforming extension type provided for an unlimited number of related, multidimensional images, which might not have the same dimensionality or binary format, to be stored in the same FITS file (Ponz *et al.* 1994). The third, and arguably most important, conforming extension was defined by Cotton *et al.* (1995).

This "binary tables convention" was first conceived in about 1984, prototyped at NRAO, and finally negotiated into a more general agreement by 1991. This extension conveys data that are logically organized in a table, an ordered collection of rows and columns. Each row has the same length and each column has the same binary type. However, different columns may be of different binary type including bit arrays, character strings, 8-, 16-, and 32-bit integers, and 32- and 64-bit IEEE floating-point numbers. The big-endian byte order of FITS binary data is retained for all binary table data. Furthermore, a column may be defined to contain an array of numbers of arbitrary size in each row. This extension thereby encompasses all of the previous data forms with the only differences being in the header keywords of the HDU.

Grosbøl *et al.* (1988) also described a change in the FITS standard to allow for data blocking. Up to 10 logical records are allowed to be stored in a single physical record. All physical records within a file must be the same length except the last one which may contain a smaller number of logical records. When the transport medium is a disk file or a transmission over a computer network or the Internet, the meaning of physical record becomes somewhat unclear but does not matter. Nonetheless, the logical record remains 2880 bytes and the only padding of data allowed are the blank fill at the end of the last header record and the zero fill at the end of the last binary data record within each HDU.

Decisions requiring all NASA missions to provide science data products in FITS format led the NASA/Science Office of Standards and Technology (NOST) to establish the FITS Support Office in 1990 to assist NASA missions to understand and implement that format. NOST also commissioned the first of the FITS Technical Panels whose task was to recast the published FITS papers into a form acceptable as an official NASA standard. That process has produced a number of standards in the period from 1990 to 2000. The last (so far), NOST 100-2.0 (Hanisch *et al.* 2001), has been adopted by the IAU FITS Working Group as the official statement of the FITS standard.

Although FITS is a format for data transport, the speed of modern computers allows it to be used as the main internal format in data analysis software packages. The overhead of scanning a full header for a needed parameter rather than having the parameters in a fixed binary structure and the need to swap bytes on some machines is no longer a barrier. To assist in this process, a significant collection of software tools has been developed at NASA Goddard in the high energy astrophysics group and made available to the astronomy community; see, for example, Pence (1992 and 1999). This package (`FITSIO`) together with other FITS-based public-domain packages (referenced from `http://heasarc.gsfc.nasa.gov/docs/software/`) have become mainstays in astronomical research and software development.

4. Successes and failures

FITS has been an unparalleled success. It has enabled countless bytes of data to be transmitted from one computer architecture, observatory, astronomer, and software system to another with every byte being correctly assigned to the proper image or table row and column. It is used by essentially all astronomical observatories, scientists, and software systems either as their fundamental data format or, at least, as an available and understood format. So far as this author knows, there are no other fields of human endeavor which have attained anything like this level of data interchange. Nonetheless, we have achieved only mixed success in exchanging the *meaning* of those bytes we have so accurately transmitted.

Wells (2000) provides a good summary of the many reasons for this success. I will reiterate some of them with perhaps a slightly different view. The basic design negotiations occurred between two, and only two, designers both of whom represented major data producing organizations whose basic role in life is to distribute data to astronomers from other institutions. The two designers were both in a position to implement any agreement in important portions of their employer's software systems. They represented two different fields of astronomy and were able to define a generalization of their prior practices. The encouragement and technical advice provided by the institutions' management and by Barry Clark and the other VLA scientists who participated in some of the meetings were also important.

The initial proposal was appreciated at some level when it first appeared and its adoption was encouraged by a variety of people. Certainly the need for some sort of transport format was clear and basic FITS was able to be read and written in very simplified ways. Harten committed the Westerbork Synthesis Radio Telescope to the format and Rudi Albrecht provided several opportunities for its promotion. Hauck and Westerhout encouraged its adoption by the IAU before it was fully implemented in most places. At the same time, the initial proposal was under-appreciated. On the surface, it was "only" a transport format and it was unlike any that previously existed so that everyone would have to write software to read and write it.

In fact, FITS encourages a particular model of data and most software systems designed since 1979 have been profoundly affected by that way of thinking about data. Had this fact been appreciated then, as it is today, I wonder if FITS or any other format could have been so widely adopted. In 1979, a FITS negotiation of very broad impact required only a few days. Now, the negotiation for correcting DATE keywords for our "Y2K" error[1] required approximately two years despite the fact that the basic answer was obvious to everyone. The problem now is that both FITS and systems of time measurement are fundamental *internal* parts of several institution's scientific software systems and any change will obviously have a noticeable *internal* impact.

Advancements in the FITS format have been helped by the creation in May 1991 of a news group called `sci.astro.fits` with a mirrored e-mail exploder called `fitsbits`. Additional e-mail exploders for specialized interests were also created as needed. These exploders have certainly enabled interested parties to remain current with, and contribute to, public on-going discussions of FITS issues. They have also enabled people to ask for, and receive, help with FITS usage and application problems. Discussions on

[1]The original FITS specified DATE strings as 'DD/MM/YY', a form that is unable to define the century. When FITS survived into the next century, a correction was essential. The new strings are in the form 'CCYY-MM-DDThh:mm:ss[.sss...]'.

these exploders are capable of becoming quite voluminous and have even achieved some sort of consensus occasionally. Frequently, however, matters are discussed briefly and then apparently dropped.

Several of the fundamental design decisions have also been instrumental in the success of FITS. Probably the most important was the decision that no addition to the format should make existing data sets obsolete. The "once FITS, always FITS" rule has meant that archives of data remain readable by modern FITS software without need for updating and format conversion. FITS reading and writing software may remain static, so long as the needs of that software remain static. In 1979, software was "free" and computers were expensive. Now, super computers are essentially free, but good software has become extremely expensive. If the format remains stable, software costs are minimized. Furthermore, if the format were to undergo a major revision, many institutions might take that opportunity to select an alternative format.

The decisions to allow new keywords and "special records" following the defined ones were also major sources of flexibility and longevity for the FITS format. The generalized extensions agreement (Grosbøl *et al.* 1988) provided a framework within which complex agreements over data structures could be negotiated, but it carefully did not rule out new types of special records to follow the standard extensions.

FITS is a syntactic standard, not a semantic standard. It has been very successful therefore in conveying the form or structure of the data, but it has had notably less success in conveying the meaning of the data. Generalized FITS reading programs can read almost any FITS binary table and can convert, with limitations, the bytes into the internal format of the host software system. The software may then use generalized routines to display for the user the names and data contents of the table columns. However, without additional negotiated conventions, that software cannot know that, for example, Column 4 contains calibration data to be applied to the image in Column 7 whose coordinates are given in Columns 5 and 6.

Wells *et al.* (1981) suggested a variety of keywords for defining coordinates and mandated the use of International System of Units (*e.g.*, meters, kilograms, seconds) for units. Despite the IAU endorsement, these suggestions were widely ignored in favor of "more natural" units and locally invented keywords which duplicate the meaning of existing keywords.

Another impediment to data interchange arose from the very flexibility of FITS. It is more difficult, time consuming, and expensive to write software to handle a wide range of possible input data even when the analysis algorithms might be capable of performing interesting operations on that range of data. Therefore, organizations often choose to write programs for a limited subset of the FITS capabilities. A great many programs have been written, for example, to read only two-dimensional images since that is all that the software designers considered that they would need to analyze. These programs were not written to accommodate the more general data representations given in Wells *et al.* (1981) in which "degenerate" (1-pixel) axes may be used to convey additional coordinate information. Thus, incompatibility arises between software systems that are aware, or not, of such representations and interoperability is compromised. Since software is expensive, and updating legacy applications is prone to unearthing or causing other problems, advances in the FITS standard must attempt to accommodate the varying levels of generality. This makes negotiations much more difficult, often resulting in compromises that are less than ideal.

There are criticisms of FITS that should be mentioned here, although many of these suggestions would tend, in my opinion, to diminish the simplicity and predictability which have been among the reasons for FITS' acceptance. (In fact, the strongest complaint heard circa 1980 was that binary data were too hard to read and that we should have provided a character form for the primary image data.) Some suggestions would even cause uncorrected FITS readers to misinterpret the headers, rather than simply failing to understand new constructs and keywords. This must be avoided in order to retain the "once FITS, always FITS" rule.

Although the logical association of FITS HDUs is explicitly conveyed by their presence within a single file, no additional information about that association is defined. Advanced software systems frequently make good use of hierarchal data structures, but they have no way to represent those structures in standard FITS. A set of the coordinates of an image may be viewed as an "object" in modern software parlance. Several such objects may, within the proposals discussed below, be used to describe the same image. But no "inheritance" from one object to another has been defined even within the same HDU, let alone between HDUs. The limitation of FITS keywords to eight, upper-case characters frequently frustrates designers of new concepts. However, the discipline that rule enforces has caused concepts to be more carefully considered and then limited them to manageable dimensions. The requirement to specify the length of the data in advance (in the header) poses significant complications for data acquisition systems.

5. World Coordinate Systems

World coordinates are the coordinates that serve to locate a measurement in some multi-dimensional parameter space. They include, for example, a measurable quantity such as the frequency or wavelength associated with each point in a spectrum or the longitude and latitude in a spherical coordinate system which define a direction in space. World coordinates may also include enumerations, such as "Stokes parameters", which do not form a normal image axis since interpolation along such axes is not meaningful.

Wells *et al.* (1981) recognized the need for world coordinate system (WCS) keywords and provided keywords for each axis of the image to specify coordinate type and a reference point for which the pixel coordinate, a coordinate value, and an increment were given. An undefined "rotation" parameter was also provided for each axis. These descriptions were kept simple so that controversy over coordinate specification would not interfere with adoption of the basic structures of the format. While participating in the development of the AIPS software package of the NRAO—see the chapter on AIPS in this volume—Greisen (1983, 1986) found it necessary to supply additional details to the coordinate definitions for spectral and celestial coordinates. These specifications were widely used in radio astronomy and some X-ray, optical, and infra-red projects also adopted them.

The negotiations on WCS have been the most protracted and complex negotiations in the history of FITS. They began with a NASA-sponsored conference held in January 1988 at the NRAO in Charlottesville. That conference recommended that a general WCS standard be based on the AIPS specifications where possible and extended to support a more general approach to handling scaling and skew (Hanisch & Wells 1988). Several variations on the notations suggested in that meeting made their way into software developments at the Space Telescope Science Institute (STScI) for data from the Hubble Space Telescope and other NASA missions and at the National Optical Astronomy Observatory (NOAO) in the IRAF software package. In response to a discussion held at the 1992 ADASS meeting, Greisen and Mark Calabretta (Australia Telescope National Facility) prepared a draft standard by December 1992 and presented it at the June 1993 AAS meeting in Berkeley (1993). Discussions with Doug Tody (NOAO) at that time led to a new version of the proposal, distributed by August 1993, which changed some of the notations of 1988 (Greisen & Calabretta 1995). A new version was offered in 1996 that added WCS keywords for binary table extensions and a method for converting real images (*e.g.*, with warps) into the ideal projective geometries previously described.

Why, after some eight years, had the community not reached an agreement? Previous negotiations had involved individuals from different areas of astronomy with the clout to implement their proposals. By 1995, Calabretta had written `wcslib`, a portable subroutine package implementing the proposal which was (and is) widely used. Despite this, there was a perception that the proposal solved the WCS problems of radio astronomy, but not of other kinds of instruments. Groups that should have invested effort to correct this situation did not. Project managers were satisfied if their software could understand its own WCS and saw little need for the expense of creating and implementing higher levels of standardization. Many of these projects were able, with some pain, to work around the lack of standards when reading a foreign WCS. Furthermore, there was, and still is, little personal reward or recognition for major efforts toward standardization and such efforts must be undertaken above and beyond other responsibilities.

These issues continued to hinder the WCS negotiations. WCS was discussed at some length at the ADASS meetings of 1997 and 1998 with the results presented in 1999 by Calabretta & Greisen (2000). The participants at the 1999 meeting voted that the papers should now be presented to the regional FITS Committees for a vote. But that did not happen. Finally on 30 June 2001, a significant generalization of the three papers (the paper had been split and spectral axes had been added) was suggested by Francisco Valdes, Doug Tody, and Lindsey Davis of NOAO. Their proposal resulted in a separation of instrumental peculiarities into a Paper IV while Papers I–III would be concerned with ideal coordinates. That proposal was presented at the 2001 ADASS and additional compromises were reached. But there were serious differences of opinion remaining which were only resolved with the skillful mediations of Bob Hanisch (STScI) and by simply recognizing the validity of both of the competing nomenclatures.

The WCS negotiations have been exhausting. Several groups helped in areas of their expertise, particularly tables and representations of units, and a very few individuals were consistently helpful and supportive. However, the lengthy periods of inactivity, the apparent inattention of knowledgeable individuals, the perceived antagonism, and lengthy arguments over issues that may be considered matters of taste make undertaking a new FITS proposal quite daunting. And that neglects the very real difficulties associated with a complex subject like world coordinates. The long period used by this negotiation also allowed WCS solutions that differed from the early proposals to appear in a great many FITS files, thereby achieving a need to be supported without having any community agreement.

All this being said and despite a few committee-like compromises, the resulting papers are far more comprehensive, accurate, and readable than the drafts of even one or two years ago. Although the main portions resemble the earliest drafts, significant details have been added, strengthened, and corrected in order to insure functional implementation. The first two WCS papers have now passed the North American FITS Committee and have been submitted (Greisen & Calabretta 2002; Calabretta & Greisen 2002). Paper III on spectral coordinates is in a nearly final draft form (Greisen *et al.* 2003). Paper IV on distortion correction is in preparation.

The WCS experience need not be repeated. In the new Virtual Observatory framework, international agreement was reached on a representation of tabular data in XML (drawing upon the vast body of FITS experience) in a matter of months (Ochsenbein *et al.* 2002). The process can work well – and quickly – when there is a will to succeed.

Our community needs to adopt a more aggressive and inclusive process for standards development.

6. Scientific achievements of FITS

This section is comparable to a section on the intellectual achievements of a spoken language[2] or even a screwdriver. Like them, FITS is a basic tool that makes all sorts of things possible and like them, their use is never mentioned. Because FITS is nearly ubiquitous, a user of one telescope may choose to reduce the data not with the software provided for that telescope, but with software provided in connection with some other, usually similar, telescope. FITS allows convenience, familiarity, and availability of algorithms to determine where and how a scientist analyzes the data. Furthermore, it determines when the analysis may take place. If data are archived in FITS, the very stability of the format guarantees that the data are accessible to the scientist with a new scientific question, perspective, or algorithm long into the future. FITS is also widely used to serve images over the web (*e.g.,* Condon *et al.* 1998) and will be a major tool employed by the Virtual Observatory (*e.g.,* Szalay 2001).

FITS enables the scientist to observe with different instruments over a wide range of wavelengths such as X-ray, optical, infra-red, and radio and then to bring all the images together in order to learn much more about the physics of the objects. Multi-wavelength projects now seem routine or even obligatory, whereas they were rarely carried out when FITS was first invented. So that the images may be aligned, an accurate WCS must be available for each image. Until recently, such information was not always available in widely understood keywords on all FITS files. That situation is improving.

One early experiment, conducted around the time FITS was first accepted by the IAU, is illustrated in Fig. 1. The twin, wide-angle-tail radio galaxy 3C75 was observed in 1983 by Owen *et al.* (1985) using the Very Large Array at 20- and 6-cm wavelength in multiple configurations. Those data were calibrated and imaged with the NRAO software of that era and written as FITS images. Also in 1983, CCD images were made at R and B bands using the NOAO 0.9-m telescope and written on FITS tapes. The images were then processed in the same software system to compare the appearance of the galaxy at these four frequencies. The optical data had

[2]FITS does well in defining the syntax (grammar) but not so well on the semantics (vocabulary). Non-standard keywords are comparable to different dialects, while new standard keywords are like new words which have to be learned.

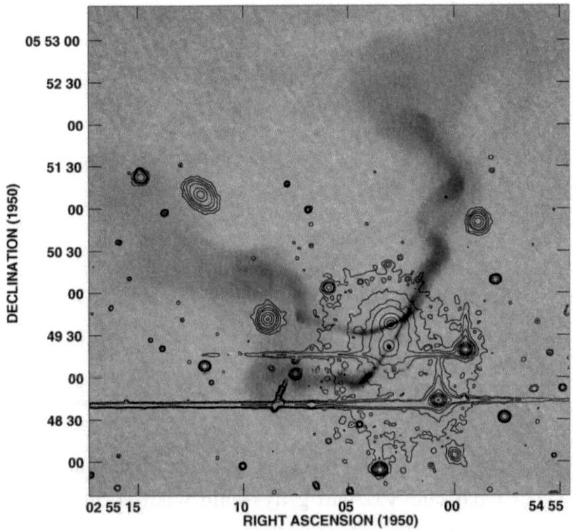

Figure 1. 3C75 observed in B band (contours) at the NOAO 0.9-m telescope by Frazer N. Owen and Richard A. White in 1983. The gray-scale image was also made in 1983 using the NRAO VLA at 6-cm wavelength by Owen *et al.* (1985).

to have the WCS parameters determined by traditional means since they were not recorded on the FITS tape (and probably were unavailable at the time the tape was written). The WCS parameters of the radio image are well known due to the nature of the instrument. The two central dots, seen in both radio and optical, are the twin nuclei of the central galaxy in the cluster of galaxies Abell 400. The radio jets are bent, possibly by the motion of the nuclei through the hot gas in the cluster. On the right side of the image, the jets appear to interact and possibly are wrapped around each other. The diffuse stellar light of the galaxy may be seen surrounding the two nuclei. The two objects with spikes are stars; several other galaxies and foreground stars may also be seen in the contours.

There are so many examples of this multi-wavelength astrophysics that any selection done here should be regarded as random and certainly neglects the most important such papers. Nonetheless, there are several items worth mentioning. The web site of the NRAO is beginning to contain a gallery of images at `http://www.nrao.edu/imagegallery`. A number of the images contain optical as well as radio data, particularly the pages on radio galaxies and neutral hydrogen in galaxies. See also Hibbard *et al.* for the latter. Bauer *al.* (2000) have correlated ROSAT and NRAO VLA Sky Survey source lists and then identified a large number of objects with them optically. Falcke *et al.* (1998) have found striking high-resolution correlations between the Hα and radio structures in Seyfert galaxies using Hubble Space Telescope and VLA imagery. Blanton *et al.* (2001) use Chandra X-

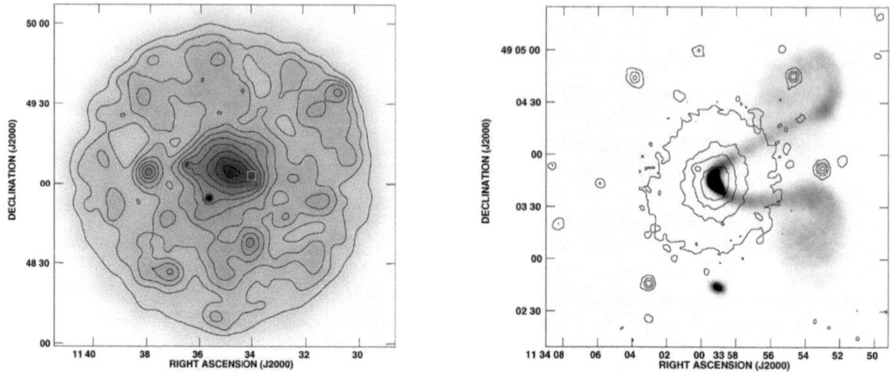

Figure 2. Left: galaxy cluster Abell 1314 observed in X-rays with ROSAT; the small white box shows the location of the field containing IC 708. Right: the radio galaxy IC 708 observed at 4.5 GHz with the VLA in gray-scale with contours of the optical image from the POSS.

ray and VLA radio images of a cooling-flow cluster of galaxies to show a coincidence in the X-ray and radio central core sources and a correlation between the radio lobes and holes in the X-ray emission. They write "The data are consistent with the radio source displacing and compressing, and at the same time being confined by, the X-ray gas." Chu *et al.* (2001) have combined Chandra X-ray and Hubble Space Telescope Hα imagery to analyze the densities and temperatures in NGC 6543, the Cat's Eye Nebula.

Abell 1314, a cluster of galaxies located at a redshift of 0.0338, provides another example. It contains large diameter intra-cluster X-ray emission seen in the ROSAT image[3] in the left half of Fig. 2. The cluster contains IC 711, a radio galaxy with an exceptionally long tail extending over 600 kpc in projection, and the radio galaxy IC 708, illustrated in the right half of Fig. 2. The gray-scale image is of the radio source at 4.535 GHz made with the Very Large Array at a resolution of about 5 arc seconds (Clarke & Vogt 2002).

The optical image shown in contours is taken from the POSS[4] and shows a diffuse galaxy coincident with the central radio source. The unusual radio

[3]This research has made use of data obtained through the High Energy Astrophysics Science Archive Research Center Online Service, provided by the NASA/Goddard Space Flight Center.

[4]Based on photographic data of the National Geographic Society – Palomar Observatory Sky Survey (NGS-POSS) obtained using the Oschin Telescope on Palomar Mountain. The NGS-POSS was funded by a grant from the National Geographic Society to the California Institute of Technology. The plates were processed into the present compressed digital form with their permission. The Digitized Sky Survey was produced at the Space Telescope Science Institute under US Government grant NAG W-2166.

structure of IC 708 may be related to its gravitational interaction with the nearby galaxy IC 709 and the cluster center. In this scenario, our line of sight lies close to the orbital plane of the radio source. The compact radio source south of IC 708 has no obvious counterpart on the optical image and is likely to be an unrelated background source.

7. Summary

FITS, the Flexible Image Transport System, provides methods by which astronomers can exchange their data. It is efficient, straightforward, unambiguous, interpretable, flexible, and powerful. It achieves these
(1) by using binary recording of the image and tabular data at user-selected accuracy,
(2) by specifying fixed logical record lengths, industry-standard coding of characters and binary data, and a simple general structure,
(3) by requiring a minimal and accurate description of the data records,
(4) by expressing all header parameters in ASCII text which can be read by humans as well as computers,
(5) by providing a general set of keywords,
(6) by making a general form for extensions and defining the extremely general binary tables extension, and
(7) by allowing the creation of new keywords in the header and new record types following the main HDU. FITS is used throughout the astronomical community and has been adopted as an IAU standard.

Acknowledgements

The author is grateful to Bob Hanisch, Mark Calabretta, and Steve Allen (UCO Lick) for comments on this manuscript and to Frazer Owen and Tracy Clarke (both NRAO) for providing images, some prior to publication.

References

1. Bauer, F.E., Condon, J.J., Thuan, T.X. & Broderick, J.J. 2000, RBSC-NVSS sample. I. radio and optical identifications of a complete sample of 1556 bright X-ray sources, *Ap. J. Suppl. Series* **129**, 547.
2. Blanton, E.L., Sarazin, C.L., McNamara, B.R. & Wise, M.W. 2001, Chandra observation of the radio source/X-ray gas interaction in the cooling flow cluster Abell 2052, *Ap. J.* **558**, L15.
3. Calabretta, M.R. & Greisen, E.W. 2000, Representations of World Coordinates in FITS, *Astronomical Data Analysis Software and Systems – IX*, Manset, N., Veillet, C. Crabtree, D. Eds., *A. S. P. Conference Series* **216**, 571.
4. Calabretta, M.R. & Greisen, E.W. 2002, Representations of Celestial Coordinates in FITS, submitted to *Astr. & Astrophys.*, astro-ph/0207413.
5. Chu, Y.-H., Guerrero, M.A., Gruendl, R.A., Williams, R.M. & Kaler, J.B. 2001, Chandra reveals the X-ray glint in the Cat's Eye, *Ap. J.* **553**, L69.

6. Clarke, T.C. & Vogt, C. 2002, in preparation.
7. Condon, J.J., Cotton, W.D., Greisen, E.W., Yin, Q.F., Perley, R.A., Taylor, G.B. & Broderick, J.J. 1998, The NRAO VLA Sky Survey, *A.J.* **115**, 1693.
8. Cotton, W.D., Tody, D. & Pence, W.D. 1995, Binary table extension to FITS, *Astr. & Astrophys. Suppl. Ser.* **113**, 159.
9. Falcke, H., Wilson, A.S. & Simpson, C. 1998, Hubble Space Telescope and VLA observations of Seyfert 2 galaxies: the relationship between radio ejecta and the narrow-line region, *Ap. J.* **502**, 199.
10. Greisen, E.W. 1983, Non-linear Coordinate Systems in AIPS, *AIPS Memo Series* **27**, National Radio Astronomy Observatory, Charlottesville, Virginia. Also appeared in 1984, Eds. Albrecht, R. & Capaccioli, M., *IAU Working Group on Astronomical Image Processing*, Circular Number 10, Space Telescope Science Institute, Baltimore, Maryland.
11. Greisen, E.W. 1986, Additional Non-linear Coordinates, *AIPS Memo Series* **46**, National Radio Astronomy Observatory, Charlottesville, Virginia.
12. Greisen, E.W. & Calabretta, M.R. 1993, Representations of Celestial Coordinates in FITS, *B.A.A.S.* **25, No. 2**, 808.
13. Greisen, E.W. & Calabretta, M.R. 1995, Representations of Celestial Coordinates in FITS, *Astronomical Data Analysis Software and Systems – IV*, Shaw, R. A., Payne, H. E., Hayes, J. J. E. Eds., *A. S. P. Conference Series* **77**, 233.
14. Greisen, E.W. & Calabretta, M.R. 2002, Representations of World Coordinates in FITS, submitted to *Astr. & Astrophys.*, astro-ph/0207407.
15. Greisen, E.W. & Harten, R.H. 1981, An extension of FITS for groups of small arrays of data, *Astr. & Astrophys. Suppl. Ser.* **44**, 371.
16. Greisen, E.W., Valdes, F.G., Calabretta, M.R. & Allen, S.L. 2003, Representations of spectral coordinates in FITS, to be submitted to *Astr. & Astrophys.*, see `http://www.aoc.nrao.edu/~egreisen/`.
17. Grosbøl, P., Harten, R.H., Greisen, E.W. & Wells, D.C. 1988, Generalized extensions and blocking factors for FITS, *Astr. & Astrophys. Suppl. Ser.* **73**, 359.
18. Hanisch, R.J., Farris, A., Greisen, E.W., Pence, W.D., Schlesinger, B.M., Teuben, P.J., Thompson, R.W. & Warnock III, A. 2001, Definition of the Flexible Image Transport System, *Astr. & Astrophys.* **376**, 359.
19. Hanisch, R.J. & Wells, D.C. 1988, World coordinate systems representations within the fits format, notes from a meeting sponsored by the National Aeronautics and Space Administration, Code EZ, Charlottesville, Virginia. Located at `http://www.cv.nrao.edu/fits/wcs/wcs88.ps.Z`.
20. Harten, R.H., Grosbøl, P., Greisen, E.W. & Wells, D.C. 1988, The FITS tables extension, *Astr. & Astrophys. Suppl. Ser.* **73**, 365.
21. Hibbard, J.E., van Gorkom, J.H., Rupen, M.P. & Schiminovich, D. 2001, An HI rogues gallery, *Gas and Galaxy Evolution*, Hibbard, J. E., Rupen, M. P, van Gorkom, J. H. Eds., *A. S. P. Conference Series* **240**, 659..
22. IAU 1982, Commission Resolution C1, *Transactions of the International Astronomical Union* **XVIIIB**, 45.
23. Ochsenbein, F., Williams, R., Davenhall, C., Durand, D., Fernique, P., Giaretta, D., Hanisch, R., McGlynn, T., Szalay, A. & Wicenec, A. 2002, *Toward an International Virtual Observatory*, ESO Astrophysics Symposia, in press.
24. Owen, F.N., O'Dea, C.P., Inoue, M. & Eilek, J.A. 1985, VLA observations of the multiple jet galaxy 3C 75, *Ap. J.* **294**, L85.
25. Pence, W.D. 1992, FITSIO and FITS file utility software, *Astronomical Data Analysis Software and Systems – I*, Worrall, D. M. , Biemesderfer, C., Barnes, J. Eds., *A. S. P. Conference Series* **25**, 22.
26. Pence, W.D. 1999, CFITSIO, v2.0: a new full-featured data interface, *Astronomical Data Analysis Software and Systems – VIII*, Mehringer, D. M., Plante, R. L., Roberts, D. A. Eds., *A. S. P. Conference Series* **172**, 487.
27. Ponz, J.D., Thompson, R.W. & Muñoz, J.R. 1994, The FITS Image extension,

Astr. & Astrophys. Suppl. Ser. **105**, 53.
28. Szalay, A. S. 2001, The Virtual Observatory, *Astronomical Data Analysis Software and Systems – X*, Harnden, F. R., Primini, F. A., Payne, H. E. Eds., *A. S. P. Conference Series* **238**, 3.
29. Wells, D.C. 2000, The FITS experience: lessons learned, *Information Handling in Astronomy*, Heck, A. Ed., Kluwer Academic Publishers, Dordrecht, 65.
30. Wells, D.C. & Greisen, E.W. 1979, FITS: a Flexible Image Transport System, *Image Processing in Astronomy*, Eds. Sedmak, G., Capaccioli, M., Allen, R. J., Osservatorio astronomico di Trieste, Trieste, 445.
31. Wells, D.C., Greisen, E.W. & Harten, R.H. 1981, FITS: a Flexible Image Transport System, *Astr. & Astrophys. Suppl. Ser.* **44**, 363.

THE MUNICH IMAGE DATA ANALYSIS SYSTEM

K. BANSE
European Southern Observatory
Karl-Schwarzschild-Straße 2
D-85748 Garching, Germany
kbanse@eso.org

Abstract. The history of MIDAS and its evolution over the past 20 years is described. This includes the many upgrades of the system necessary to follow the changes in technology during that time: from proprietary host systems to Linux, from TV-like displays to desktops with window systems. Beyond giving the technical facts, emphasis is placed on showing also the people who made all that happen.

1. The Planning

ESO has a long experience with astronomical image processing systems. Already since 1975 an HP 1000 series based image processing system, the Image Handling And Processing system, `IHAP`, was developed by the late Frank Middelburg at ESO's Telescope Project Division on the premises of CERN in Geneva, Switzerland, and heavily used by astronomers who had obtained data with ESO telescopes in Chile[1]

Data reduction was done in quite a different way than we are used to these days: Only the bigger institutes had the money and resources to set up data reduction centers. Consequently, astronomers who obtained observing time at ESO telescopes would also get support for travelling to Geneva after their observation run (and to Garching later on) to reduce their data using the ESO facilities and astronomical data reduction software.

With the arrival of the CCDs in the late 70's it was clear that these new detectors, which produced digital data right away, would also require more powerful data reduction systems. Also around this time the first "midi"

[1]For a detailed history of IHAP, refer to the chapter by Grøbol & Biereichel in this volume.

A. Heck (ed.), Information Handling in Astronomy – Historical Vistas, 89-107.

computers appeared which used a 32bit address space and supported virtual memory. At ESO, a study for the data reduction requirements for the 80's was written in the summer of 1979 which emphasized the need to move to new hardware and to a new generation of image processing software. The eventual decision was to go for VAX 11/780 computers from DEC and sophisticated interactive image display devices.

ESO was not alone in looking for new avenues for the reduction of the great amount of data which was expected to be produced by the new generation of detectors for astronomical instrumentation. Also ESA (for `FIPS`, the software to test the performance of the Faint Object camera) and the UK were actively planning new systems – and they also chose the VAX 11/780 systems like ESO. The UK project, STARLINK, was the most ambitious one with the goal of providing different centers all over Great Britain with VAXes, and much more advanced in the design of the new system than us.

Thus in the 80's the 32bit VAX/VMS systems were the dominant computer family used in the astronomical community in Europe (and in the US as well).

For the specifications of the software requirements ESO could profit from the experiences made so far with the existing IHAP system which, in 1979, was heavily utilized in La Silla to support the observations and also regularly used by astronomers visiting ESO in Geneva. So, the main points for the new design were:

- Like IHAP, the new system should be a command driven system geared towards the interactive user sitting in front of a terminal. It should be straight forward to combine commands to build up individual "macros" which can be executed in "batch" mode, a feature which had been appreciated a lot by the users of IHAP.
- The system should be easy to use with extensive help available on-line.
- The system should be compatible with the other major astronomical image processing systems currently developed, *i.e.* STARLINK and FIPS, the Faint-Object Camera Image Processing System of ESTEC.
- The IHAP system was perfectly adapted to the HP 1000 family of computers and optimally tuned for the relevant host system, thus the performance was truly amazing. However, the slightest change in the operating system, *e.g.* a new Fortran compiler or upgraded file management would require extensive modifications to IHAP.
  So, unlike IHAP, the new system should be portable in the sense that it would run on any computer of the same family as the one where it was developed, independent of the version of the operating system and the peripherals connected.

- The new system should be exportable and supported as an ESO service to its community.
- Implementation of applications for the new system should be made easy and efficient to encourage astronomers as well as software engineers to integrate their programs.
- The system should be very modular with the possibility to build a bare bone, minimum system quickly and adding applications on top.
- World coordinates should be supported consistently in the new system.
- Similar to IHAP (with the Image Disector Scanner, `IDS`) one should aim for integrating data acquisition and reduction in one system.
- Make all code public and easily modifiable by the users.

2. The design phase

In 1979 Phil Crane, an American astronomer who had joined ESO already in Geneva, became project manager for the new image processing system and began building up an Image Processing Group, `IPG` at ESO to design and develop that system. The first staff member to join Phil's group was me, this was my second major project at ESO after having worked on the control system for the ESO measuring machines before in Geneva. And the design phase of the new system took off, accompanied by extensive discussions with the STARLINK project and the group at ESTEC working on FIPS.

The STARLINK group had already a design concept[2] which fulfilled quite a few of the requirements of ESO, also they already had access to a VAX and had gained experience with VMS while writing a first prototype of the low level system to test the basic data structures. We took this prototype software as a template and soon after ESO's move to Munich in the summer of 1980, the first VAX 11/780 was installed and the actual development work could begin.

Several ideas from the STARLINK prototype system were taken over, as well as many terms (like the file type `BDF = bulk data frame` for images), others abandoned in favour of a more compact and unified system, and code inspection of the software helped greatly in getting up to speed with the VAX/VMS environment. This was not the most desirable approach from a programmers point of view – we would have preferred to start from scratch and create something totally new ...

[2] After having reviewed a draft of this article, Rein Warmels made me aware of the fact that STARLINK in turn got many ideas for their system from `GIPSY`, the image processing system developed at Groningen in the Netherlands already since 1977. Thus credit should also be given to GIPSY, however, the IPG didn't have contacts with the GIPSY group at that time.

However, that way we saved probably several months in the design and development phase of MIDAS, and time was what we didn't have[3]!

We finally settled on a name for the new system which Phil came up with: `MIDAS`, the Munich Image Data Analysis System. This name was later changed to ESO-MIDAS for copyright reasons, since there did exist already a financial software package with the name of Midas.

3. Putting the Pieces together

In 1981 the design for MIDAS converged to a stable state and the implementation phase could begin.

This year also the hardware environment for MIDAS could be put together; due to Phil's efforts and talent of persuasion, ESO management approved all our wishes and in early 1982 the following configuration was completed:
– The main computers were two VAX 11/780 linked together via DECnet.
– The VAXes were equipped with 3.5MB (resp. 4 MB) memory, 1.2 GB (688 MB) of disk space and 2 (1) tape drives with 800/1600 bpi (1600/6250 bpi) density.

The image display systems used for MIDAS were 3 Gould-DeAnza IP-8500 systems. Each system was equipped with an array processor working on entire frames of 512 × 512 pixels at video rates, and supported two user stations (displays). Each such station had 4 image and 1 overlay channels of 512 × 512 pixels of 8 bits, as well as an alpha-numerics memory of 20*80 characters and two independent cursors. Images of 512 × 512 size were considered big images then, and the processing of such images was pushing the computer systems to the limit ...

Two of the IP8500 systems were connected to one VAX, the third to the second VAX. A Dicomed image recorder served as a high-resolution hardcopy device producing publication-quality output on regular roll film which was then developed outside. In addition, a Versatec plotter was connected to each VAX, and one HP pen-plotter was available as well. In total, ESO offered to in-house users and Visiting Astronomers 6 image processing workstations, each consisting of a DeAnza display, a VT 100 terminal for interaction with MIDAS and character I/O and an HP 2648 terminal for graphical output.

Nowadays, we smile about computers with 4MB memory – but in those times, this was a *"Mercedes class"* installation, worth millions of

[3]The fruitful collaboration with STARLINK didn't last long, though: A year later, they came up with a new, drastically changed, system design, whereas we decided to continue with our original ideas and design.

Deutschmarks and for which ESO was envied by most institutes in Europe and beyond!

In the meantime Charlie Ounnas, Preben Grosbøl and Daniel Ponz had joined the Image Processing Group. Charlie came from Nice Observatory and would work on graphics software of MIDAS. Preben, a Danish astronomer came to ESO already in Geneva, then as a fellow, and had been the driving force behind the table system which had been installed in IHAP. He would look into all issues concerning the interaction of MIDAS and IHAP. Daniel had worked before at ESA's IUE Observatory in Villafranca del Castillo, Spain and his job was to implement a powerful table system also in MIDAS, probably the most important subsytem of MIDAS as well as to write instrument related applications.

4. The first release

In 1982 the first version of MIDAS was officially released in-house at ESO, parallel to IHAP, and offered for data reduction; local users and Visiting Astronomers started to use it (*e.g.* for analyzing CCD data) first on an experimental basis and later on for real.
According to the design plan the following main features were implemented by then:

- MIDAS was a modular and hierarchical system, based on a limited set of low-level interfaces which provide the building blocks for everything else. These basic interfaces, written in VAX/VMS Fortran, provide access to all the MIDAS data structures:
  - images consisting of descriptors (the ancillary information) and pixels with equal sampling;
  - tables consisting of descriptors and rows and columns with unequally sampled and heterogeneous data;
  - keywords, a global data structure similar to a labeled `COMMON` in FORTRAN for communication between different applications and MIDAS[4].
- A central module, the MIDAS monitor, provides the (character-based) interface to the user and executes a limited set of basic commands directly.
  The structure and syntax of the MIDAS commands had been modeled after the DEC Command Language, `DCL`, the command language of the VMS operating system on the VAX. With commands like, *e.g.*

[4]For readers who didn't grow up with FORTRAN as the main scientific programming language: the `COMMON` construct in FORTRAN supports a set of global variables to be shared *e.g.* by the main and other functions, *i.e.* comparable to static variables in C.

`READ/TABLE`, where the command describes the general action to be performed (an English verb) and the qualifier specifies the object on which to perform that action or specific options of the action. We chose DCL because:

- that was the environment we used for all host system related functionality and with which we were familiar;
- the `command/qualifier` naming seemed rather natural;
- it integrated the MIDAS system very smoothly into the VMS environment, and thus, would be easier to learn for a new MIDAS user who uses the VAX/VMS environment in his/her daily work.

– Furthermore, the monitor serves as an interpreter for the MIDAS command language, `MCL`. This MIDAS command language was not intended to be a real programming language. Instead it should be seen as a Macro language which supports quick and easy creation of scripts, *MIDAS procedures*, combining existing commands. Thus it is possible also for the inexperienced user to build up new commands with an extended or different functionality by putting together already existing commands.
To this end, the MCL provides:

- basic control structures, like conditional branching, nested looping;
- local/global variables, parameter substitution;
- and most importantly direct access to all MIDAS data structures, so that *e.g.* the flow of control in a MIDAS procedure could depend on the value of a given table element.

All MIDAS commands which are not directly executed by the monitor are implemented as such procedures, combining basic MIDAS commands and compiled modules which are executed as subprocesses, as well as other procedures.

– The monitor and the applications operate in separate process environments, so a crashing application will not affect the rest of the system, and communicate via MIDAS `keywords`.

– The MIDAS system was written in VAX/VMS Fortran, except a few low-level file access routines which had to be coded in Macro (the VMS assembler language). All MIDAS applications would only be written in Fortran to achieve the desired level of portability between different members of the VAX computer architecture. Also, the development of applications would profit from that, because for a widely used, high level programming language debugging, support and maintenance tools were readily available.

5. MIDAS meets the World

In 1983 MIDAS finally went to Chile: The La Silla Observatory, up to then totally committed to HP computers, got it's first VAX 11/750 on the mountain and in July of that year, Charlie and I installed MIDAS there. That year, also the first outside institution, the Max-Planck Institute for Astronomy in Heidelberg, obtained MIDAS and installed it on their VAXes.

5.1. MIDAS AND THE SPACE TELESCOPE

The other main event (from a MIDAS perspective) was the arrival of the Space Telescope – European Coordination Facility, `ST-ECF` at ESO headquarters in the end of 1983; the following year most ECF staff had taken up work in Garching.

In 1984 the ECF had selected MIDAS as their data analysis host system and also supported actively the system; here one should mention especially Dietrich Baade and Michael Rosa who supported MIDAS by contributing astronomical applications as well as detailed suggestions for improvements and bug fixes.

At the Space Telescope Institute IRAF (the Image Reduction and Analysis Facility) from NOAO was chosen as host system and the Institute planned to integrate their own application package (SDAS) into that system within two years. Synchronising SDAS and MIDAS would greatly improve the collaboration and sharing of software on both sides of the Atlantic and both groups, ECF and IPG, began to work intensively on that. These efforts brought the first major change to the MIDAS system – in order to enable the integration of SDAS applications in the MIDAS environment and vice versa, we decided to move to a new, common set of basic interfaces, the `ST`andard interfaces. The first version of the low-level interfaces in MIDAS was rather VMS-specific, using special features of the VAX-FORTRAN language (which extended standard FORTRAN significantly). The table interfaces needed only an upgrade to standard FORTRAN, since SDAS did not have any. Now, these interfaces were changed to conform strictly to the ANSI FORTRAN 77 standard and their scope enlarged to satisfy the needs of sample SDAS application programs. The envisaged goal, an exchange of SDAS and MIDAS applications, was never realized (except, that SDAS implemented the table interfaces of MIDAS), but for us it was the first step in the direction of true portability.

In 1984 the ECF organized also a first Data Analysis Workshop at ESO where besides specific ST related problems also all aspects of the different data analysis systems/packages like *e.g.* MIDAS, SDAS, IRAF, FIPS were discussed.

The collaboration of the IPG with STScI lasted a few years, but then the Institute abandoned SDAS and any agreement they had with the ESO in favor of pushing IRAF as their data reduction system. Also ECF dropped MIDAS and followed in the footsteps of the Institute.

6. The Years of Change

In 1986 Phil went back to the Science Division and Preben became the new head of the IPG. The same year, Susan Lively joined the MIDAS group to help with the documentation and the MIDAS distribution. Susan was British but lived already in Munich since quite some time. François Ochsenbein had come in 1986 from Strasbourg, where he was involved with the `Simbad` database, to work on the ESO archive but also contributed to the table system, writing quite a few low-level routines for it, before he returned to France in 1992. MIDAS releases were now delivered twice a year on a regular basis, usually in May and November, to a growing number of institutes mainly in Europe but also elsewhere in the world.

6.1. IMAGE DISPLAYS

By 1985 the number of MIDAS installations had grown significantly since many institutes in Europe and the US had switched to VAX computers. Buying VAXes was possible for most sites, but purchasing an image display system was a different matter – these devices were expensive and, especially the DeAnza systems really cost money. The image display related software in MIDAS took full advantage of the capabilities of the DeAnza displays and would not work with other displays, except the Ramtek devices which ESO used with IHAP, so the MIDAS software had been modified already, when MIDAS was installed in Chile the year before.

This problem became more and more an obstacle to the portability and distribution of MIDAS (and the other image processing systems) and was a hot topic at the next Data Analysis Workshop of the ECF in 1985. So, at that occasion a working group was set up with the goal of defining a set of device independent interfaces for an abstract image display device providing the functionality needed by a fully fledged astronomical image processing system. This working group, with core members from the STScI, ECF, IPG, STARLINK and ASTRONET, but supported by quite a few other interested members of the astronomical community, first came up with a draft for an abstract model, the Image Display Interfaces, `IDI`, for astronomy.

Subsequently, prototype implementations were done by ESO (for the DeAnza) and the ASTRONET people (Fabio Pasian, Mauro Pucillo and Paolo Santin from the Trieste Observatory) for a different display (EI-

DOBRAIN 7001). Furthermore, we also upgraded all the MIDAS display software to use the new IDI interfaces. To test the concept this MIDAS release was installed in Trieste but the code was linked with the Trieste IDI-library – and it worked! After some more discussions and modifications the IDI interfaces were finalized in early 1987 and the official IDI document was published later that year. This was truly a remarkable proof of the benefits of international collaboration!

6.2. GRAPHICS SOFTWARE

Late in 1985 Rein Warmels came from the Kapteyn Institute of the University Groningen where he had been involved with GIPSY, the Groningen Image Processing System, and joined the IPG. Besides being responsible for the astronomical imaging software, he took over the graphics subsystem of MIDAS since Charlie Ounnas left the development team to become the system manager of ESO's computer setup in Garching.
Similar to the image displays we needed more portable solutions for the graphics software of MIDAS. After a survey of the already available graphics libraries in the astronomical community we settled on the Astronet Graphical Library, `AGL`, developed and maintained by Luca Fini from Florence. AGL was a portable C library supporting many different graphic devices and was extensible to add drivers for new hardware. AGL was a mature library used widely in the Italian community and thus quite well tested. We integrated AGL into MIDAS to serve as the base library for the graphics subsystem of MIDAS and with the help of Luca all graphics related modules in MIDAS were upgraded to the new library in 1986.

6.3. PORTABLE MIDAS

In the middle of the Eighties a new breed of computers, the Unix-based workstations, integrating graphics and display capabilities on the same desktop, became an attractive alternative to the VAX/VMS machines and offered a much cheaper alternative to the user stations, comprised of ASCII terminal, graphics terminal and image display used so far. In early 1985 plans for a portable version of MIDAS were discussed, where *portable* meant a version which would run on VAX/VMS as well as Unix machines. The actual work on *Portable MIDAS* began in 1986 on a MicroVAX running Unix. We had looked at other systems which were running in the Unix and VMS environments, namely IRAF and AIPS. We decided to use an approach similar to the one the AIPS group had used to achieve this portability:

- all references to the operating system inside MIDAS are done only via a specific set of interfaces, which is different for each underlying operating system;

- The MIDAS system code is rewritten in C, the ST interfaces are implemented in C and a FORTRAN 77 wrapper is provided for each interface;
- all existing MIDAS applications are converted to FORTRAN 77 with some extensions (a preprocessor is provided to transform this code to strictly standard FORTRAN 77);
- future applications can be written in FORTRAN or C (preferred).

This approach required a sizeable effort in coding and rewriting but like that we would again obtain a very open system where users could easily integrate their own software written in FORTRAN or C.

This move to Unix provided also the opportunity to change some routines of the standard interfaces library of MIDAS. Since stability and backwards compatibility had always had a very high priority, we had not modified any parameter or calling sequence to this library. But now was the one and only time to update some routines and make some necessary upgrades together with changes needed for the portability work.

The port to Unix was by far the most challenging project for the IPG, because except Preben, who through his contacts with colleagues in the Scandinavian countries had known Unix already, we had hardly any knowledge of Unix and also the language C was new territory for most of us! The VAX/VMS FORTRAN compiler had supported an extremely user friendly environment taking care of many nitty gritty details, especially in relation to character strings. The forgotten Null character at the end of C-strings would haunt us for years ...

With Carlos Guirao came in early 1988 the first "true" Unix guy who had not gone through many years of VAX/VMS experience. Carlos had worked before for Fujitsu in Spain and joined the IPG to help with the development of the portable MIDAS release.

In 1988 the first release of portable MIDAS (supporting VMS and Unix) was made available to the user community, making the IPG the first group inside ESO which had moved its main project to Unix. However, compared to the previous VMS release the first portable version of MIDAS was less stable. We gained in portability but lost robustness.

So, we started a MIDAS quality check campaign where all postdoctoral fellows at ESO (in an effort led by Dietrich) would use and test a given set of MIDAS commands and provide standardized test reports to the MIDAS group[5]. This together with the feedback from external MIDAS users enabled us to get the release back to its previous stability within the following year.

[5]To be precise, the MIDAS group and IPG were not the same, the larger IPG was also reponsible for the management and administration of the ESO main computer system in Garching as well as the Science Archive and the measuring machines.

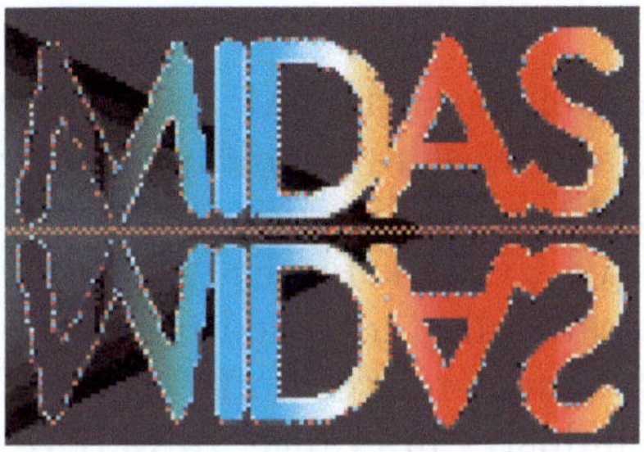

Figure 1. Midas logo

The arrival of the workstation marked also the end of the expensive peripheral image display stations as well as the need for an extra graphical terminal. Instead the window system on these workstations, first X10 then X11, provided all the features needed on a single display, so now everybody had a full image processing station on his/her desktop. For MIDAS this meant another change since the display and graphics functionality was provided in quite a different way than via the display stations before. That we had moved to the IDIs before was of great help for the necessary upgrade, because that had forced us already to cleanly separate display device dependent and independent software . It was again a very successful collaboration with the IDI gurus from Trieste, through which we brought the MIDAS display and graphics code into the workstation era. They had written a prototype image display server for X10-based on an IDI compliant environment. This code served as a template for moving on to X11 and we fully implemented the IDI environment in this context for MIDAS.

After some time of overlap with the X11-based workstations, the DeAnza display systems disappeared together with the VAXes of ESO. After extensive benchmarks we opted for SUN and workstations of this vendor became the main workhorses for MIDAS.

The decommissioning of the VAX/VMS systems posed a problem for the VMS support of MIDAS. Fortunately, help was provided by the Royal Observatory of Brussels, where Jean-Pierre de Cuypers put a lot of effort into setting up Brussels as our MIDAS test, support and certification center for VAX and Alpha systems running VMS.

7. The Golden Years

From the second half of the eighties on, MIDAS had its best eight or more years.

7.1. A MATURE SYSTEM

Michèle Peron came in 1989 from the IAP in Paris, where she had already given MIDAS courses, and was hired initially to provide expert support for the MIDAS users. When Daniel Ponz left the IPG and returned to Spain a few months later to work for ESA, Michèle took over the table subsystem. In 1990 Pascal Ballester joined the group to replace Daniel for the development of instrument-related software in MIDAS; he had worked before in the aerospace industry in France. In 1991 Resy de Ruijsscher came to take care of the task of managing and organizing the documentation and MIDAS distributions which needed urgent, professional attention after Susan had left in 1990. Resy was Dutch but was already a long time Munich resident.

We now delivered the MIDAS User's Manual in 3 big volumes and an *MIDAS Environment* document explaining the usage of the standard interfaces. Organizing these documents and keeping them up to date was becoming quite a task – the MIDAS Operating Guide was also the first ESO publication for such a purpose.

The number of MIDAS sites passed 200 comprising institutes all over the world (including the Soviet Union, China and other far away places), and the system was used for more and more projects in different institutes. The most important project was probably EXSAS, the data reduction system for the ROSAT X-ray satellite, developed by the Max-Planck Institute for Extraterrestrial Physics in Garching, which was based on MIDAS. Also the data reduction for the IUE satellite at NASA in Goddard/Washington was relying on MIDAS.

In addition to the original VAX/VMS platform MIDAS was now supporting practically all Unix systems one could buy. And also a new operating system, called Linux, had caught our attention. Linux was getting close to a version 1.0 release, offering more and more of the tools we needed (*e.g.* for us, a FORTRAN compiler or FORTRAN to C converter was essential). And so, in early 1993, we succeeded in a complete port of MIDAS to PCs running Linux. MIDAS support for Linux steadily increased over the years and we became the first group inside ESO banking on Linux as a serious host system.

The core components of MIDAS had reached a very stable and mature state so that we changed the release cycle to once a year, usually in November.

The `FITS` format became more and more important and in 1989 the IPG made an agreement with the AIPS group of NRAO to synchronize their FITS writers.

We also started with the implementation of GUIs (Graphical User Interfaces) in MIDAS, where La Silla supported us a lot and contributed GUIs for several instruments[6].

We implemented strict software configuration control for the MIDAS code (via `SCCS`) and introduced an extensive problem report system, the public domain `GNATS` package, for handling the MIDAS problem reports in an organized way. Furthermore, rigorous verification using regression test procedures and tools like *e.g. Purify, Quantify*[7] was carried out before an official release was distributed. This was the first time in ESO that software was managed in such an industry-like environment.

7.2. EXTERNAL CONTRIBUTIONS

People from other institutes would come and spend some time at ESO to integrate their software into the MIDAS system, either as executable modules using FORTRAN or C, or with the help of procedures.

- Peter Stetson integrated DAOPHOT.
- Roberto Buonnano from Rome installed ROMAFOT.
- Andrzej Kruszewski from Warsaw developed INVENTORY as a MIDAS package.
- Marguerite Pierre implemented CLOUD, a code for modelling interstellar absorption lines.
- Otmar Stahl implemented the FEROS instrument support package.
- Tino Oliva implemented IRSPEC commands.
- Alessandra Gemmo included the setup procedures for OPTOPUS.
- Andrew Young implemented PEPSYS, a package for general photometric reductions.
- Jean-Luc Starck implemented his wavelet software.
- Alex Schwarzenberg-Czerny from Warsaw converted his code for time series analysis to MIDAS.
- Almudena Prieto moved IUE and HST reduction suites into the MIDAS environment.
- Richard Hook ported the image co-addition package developed by him and Leon Lucy from IRAF/SDAS to MIDAS.
- Fionn Murtagh from ECF implemented his multivariate data analysis package.
- Also many of the astronomers working at ESO contributed to MIDAS, *e.g.* Olivier Hainaut who converted `POS1`, the Astrometry package of Richard West to MIDAS. But we should mention specifically Joe

[6]However, the GUIs in MIDAS were not heavily used (maybe they came too late) execept the one for the MIDAS help facility which really made a difference compared to the command-line help.

[7]Purify and Quantify are products of Pure Software Inc., Sunnyvale, CA.

Wampler who utilized MIDAS in all possible ways to push the performance of ESO instruments to the limit and who helped immensely in the development of the echelle reduction software.

Thus it was not absolutely necessary for ESO to have the experts for all different types of instruments in house, by integrating their specific software into a common environment we could share their knowledge. One could almost say, that MIDAS was the *Napster* for astronomical software in those times ...

7.3. THE MIDAS GROUP

The MIDAS group reached its peak in manpower with Preben leading a group of 6 people.

- Pascal was working intensely on the echelle package (and implemented the Hough function in this context) and later on put many efforts into setting up a GUI standard at ESO.
- Preben was not just managing the group but handled (and actually programmed) all FITS related matters (it helped a lot that he was the chairman of the European FITS chapter).
- Carlos worked on the operating system related interfaces, was responsible for the MIDAS installation procedures, integration and delivery of the releases.
- Besides taking care of the table system, Michèle was also involved in mathematical algorithms, and would later on write a MIDAS application for organizing the observational data, the *Data Organizer*, still the core of the ESO VLT pipelines.
- Resy was busy with the management, organization and distribution of the MIDAS documentation and the actual MIDAS releases (these were big magnetic tapes, then, shipped in specially designed, foam protected parcels) .
- Rein was responsible for the graphics subsystem in MIDAS and the astronomical image applications (*e.g.* CCD package), but still found time to become the editor of the *Midas Courier* (see below).
- My field were the overall system design, the standard interfaces, except the functions related to tables (these were Michèle's job), the display subsystem (with the IDIs), as well as the general utilities and image processing applications.

7.4. MIDAS AT ESO

Besides a regular column in the *ESO Messenger*, the official ESO publication, the MIDAS group also published from 1991 on its own newsletter, the *MIDAS Courier*, to reinforce communication with the user community. The Courier was sent out twice a year to inform the user community about all aspects of the new release and provide other information concerning the usage of MIDAS.

After many years where ESO's two image processing systems were running in parallel, MIDAS had taken the place of IHAP in Garching and also in La Silla, where it was used as image display and processing software as well as used directly for instrument pipelines, *e.g.* TIMMI. Most papers resulting from observations with ESO telescopes quoted also ESO-MIDAS as the data reduction system used[8].

The yearly ECF Data Analysis Workshops dealt mainly with the specific aspects of different image processing systems and were also the arena for hot discussions about the pros and cons of the different systems. From 1989 on, ESO and the ECF organized jointly a Data Analysis Workshop in the spring of each year. Now, the emphasis was put on the actual reduction of astronomical data independent of any specific image processing system and proceedings were published. In total, five of these yearly workshops were held at ESO until 1993, and always attracted many attendees from Europe and the US. The workshop was followed by a day devoted to MIDAS, where IPG would report about the current status and future plans for MIDAS and discuss them with the MIDAS user community.

These workshops were very successful and sparked the idea for organizing a more global conference on the subject. Thus the `ADASS` (Astronomical Data Analysis Software and Systems) conference series was born and started with the first ADASS conference held in Tucson, Arizona in 1991.

8. The Most Recent Years

In the second half of the Nineties the boundary conditions for MIDAS changed. ESO was entering the peak phase of the VLT construction, and under a new management, the software efforts at ESO were now mainly directed towards the operation of the VLT and the handling of the expected huge amounts of VLT data.

In 1993 the IPG was dissolved and superseded by a new Data Management Division. The manpower for MIDAS was reduced gradually and resources were put into the development of the VLT Data Flow System (DFS) supporting the full life cycle of VLT observations.

It was predicted, that commercial software, *e.g.* `IDL`, would take over much of the astronomical image processing needs. Also IRAF, the data reduction system from NOAO, used by the STScI (and ST-ECF here) and almost everybody else in the US, was readily available. Thus, ESO saw no

[8] Even a master thesis was written (in New Zealand) comparing MIDAS and IRAF.

need for keeping up development and full support for an in-house astronomical image processing system as an ESO service.

Consequently, MIDAS was not set up anymore as the de facto data reduction and image processing system at ESO, and new fellows and students arriving at ESO would rather use the data reduction system they had known before, thereby gradually diminishing the in-house expertise with MIDAS.

8.1. WHERE ARE WE NOW?

The MIDAS project is back at square one: after starting in 1979 with one staff member we're again down to the equivalent of one person taking care of MIDAS.

But the MIDAS project left quite a few marks and traces in the software environment we have at ESO now:

- The initial working group to come up with a data flow model was chaired by Preben and included several MIDAS group members – thus a lot of the ideas and visions from the MIDAS project found their way into the design of the Data Flow System for the VLT.
- The Data Interface Control Board, `DICB` was set up first within the MIDAS context with staff involved in the MIDAS project.
- The VLT Operations Plan was based on the experience made with MIDAS and emphasized the need to integrate tightly the data acquisition and (basic) data reduction in one common environment[9].
- The ESO archive grew out of the archive managed inside the IPG, which itself had been much influenced by the HST archive efforts (via collaboration with the ST-ECF).
- The *Data Organizer* which plays a crucial role in the VLT pipeline environment was developed as a MIDAS application helping astronomers to organize their observational data for subsequent data reduction.
- The *Reduction Block Scheduler* reused heavily existing MIDAS interfaces and the communication scheme embedded in the MIDAS monitor to schedule Reduction Blocks within a separate process.
- One of the most successful VLT instrument pipelines (UVES) is built within the MIDAS environment and also used directly by astronomers outside ESO.
- In the context of the quality control of the data obtained from the VLT MIDAS procedures are used extensively.

Currently, we still maintain MIDAS and are back to two releases per year, in February and September, to be synchronised with the VLT software

[9]I think IHAP was the very first system to integrate data acquisition and reduction with the IDS control system in La Silla.

release cycle (MIDAS is included in that package) as well as with the *Scisoft* CD distribution[10]. The February release is mainly geared towards in-house usage (VLT pipelines) but is also publicly available. The September release is the "official" MIDAS release of the year where, besides making ther release available on the Web, we also produce CDs on request. We still support VMS (via Brussels) but mainly Unix in all flavours, and Linux which seems to become the host system of choice[11]. Development is centered on upgrades and modifications of existing applications to fulfill the needs of the pipelines, data flow and quality control systems.

8.2. LOOKING BACK

Was MIDAS a success? I'm probably too close to the project to give a truly objective judgement. Instead, it seems more interesting to me to see what we had learnt from building the MIDAS system the way we did.

The decision to make MIDAS as open a system as possible, was correct and a real success in my opinion. It's absolutely necessary to provide the users with very simple and flexible means (*e.g.* standard interfaces) to integrate own code written in a standard, main-stream programming language into the data reduction system of choice. This is even more important since we are in the lucky situation of having `FITS` as a standard data format adopted by all astronomical institutes. Also, offering a high level, interpreted command language supporting direct access to all internal data structures from that language was the right way enabling also non-professional programmers to integrate applications into the system. However, instead of building up such a language on your own, it is better to choose an existing, stable interpreted language (*e.g.* Python, Perl, Tcl/Tk) with a large user base (not just the astronomical user community) to ensure support and help independent of a given institute or organisation.
The hierachical way in which one could create complex commands based on calling procedures, which were calling procedures, which were ..., and so on, offers much desired flexibility and power but leads to unnecessary complexity and most likely to performance (speed) problems. So, this language should have some translating utility, similar to a compiler, to produce something like an executable function for each high-level command of the data reduction system. This way, we would be free to either reuse that command as is in a procedure, or the related function in compiled code.

[10]This CD contains a very useful collection of astronomical software and is produced twice per year by ESO as a service to its user community.

[11]We recently ported MIDAS to Macintosh OS X – who knows, maybe there are many Macintosh aficionados out there to change this ranking ...

The powerful table subsystem which was integrated tightly into all other parts of MIDAS was a real asset and contributed a lot to the acceptance of MIDAS. In fact, it was probably the first database management system for astronomie.

Several major achievements in MIDAS (*e.g.* IDIs) were greatly supported and thus made possible through collaboration with other institutes/groups in the community. This was possible due to the open architecture of MIDAS but could have been pushed further by sharing somewhat more the "control" over the system with the user community.

Also, the approach to just put a thin layer on the host system to achieve portability is in my view more economical than creating one's own virtual operating system (maybe not as elegant, though).

And besides learning a lot, we also had a lot of fun! It was a great experience to interact with the many users of MIDAS and it was very rewarding to see that our system helped people to do science in a new or even better or faster way.

8.3. AND WHAT NOW?

Technology moves on, what was considered cutting edge a few years ago may be almost obsolete, now. *E.g.* during the initial development of MIDAS the Internet did not play the important role as it does now. Most probably, systems like MIDAS, which are strongly focussed on the local system environment (single host or network), will be superseded by globally distributed data reduction systems tightly integrated with data archives. Such *Virtual Observatories* are currently at the center of interest in the astronomical community. Together with other European astronomical centers, ESO is now leading the European effort for such a Virtual Observatory project: the Astrophysical Virtual Observatory (`AVO`). The AVO represents a collection of data centers in Europe with the goal of combining their individual resources concerning data archives, software systems and processing capabilities in order to take on the challenge of processing an ever increasing amount of astronomical data as well as untapping the wealth of locally stored data sets which are unexplored or "underexploited" because they are unconnected.

8.4. CREDITS

This chapter should not be finished without expressing my thanks and appreciation for the many people not mentioned already, who had supported MIDAS over the years, and I apologize to all the people I forgot:

- the Directors General Lodewijk Woltjer and Harry van der Laan who supported the view of MIDAS as an important and exportable ESO product and service;
- Cristian Levín who helped a lot with the GUI implementations in MIDAS and ensured the smooth running of MIDAS on any possible (and impossible) computer in La Silla;
- Hans-Ulrich Käufl who insisted on implementing his instrument pipelines in MIDAS;
- Mark Calabretta who kindly permitted the usage of his library in MIDAS for the world coordinate system;
- Herman Hensberge and Werner Verschueren who relentlessly tested and checked the MIDAS echelle code during their reduction of spectra;
- Sergio Ortolani who compared INVENTORY, DAOPHOT and ROMAFOT in every detail – in the end, he knew these packages better than anybody of the IPG (and probably better than the authors ...);
- Pat Wallace who kindly provided a master-slave scheme for efficiently using subprocesses in VMS;
- Antoine Llebaria who supported MIDAS a lot, but especially for his great drawings during the MIDAS user sessions at the Data Analysis workshops ...
- Uli Zimmermann and Carlo Izzo who made EXSAS based on MIDAS possible;
- Ralph Tremmel who installed and maintained MIDAS releases in Heidelberg from the very beginning;
- Peter Dierckx and Rinze de Roos who as system administrator and operator tuned the Unix and VAX systems for us in any possible way;
- And last but not least all the MIDAS users who provided the much needed feedback to improve and expand MIDAS!

My special thanks go to Dietrich Baade, Pascal Ballester, Preben Grosbøl and Rein Warmels for reading drafts of this paper and their many helpful comments.

AIPS, THE VLA, AND THE VLBA

E.W. GREISEN
National Radio Astronomy Observatory[†]
P.O. Box O
Socorro, NM 87801-0387, USA
egreisen@nrao.edu

Abstract. At this writing, the AIPS package has been in active development and use for over 23 years. It is still the software of choice for all phases of data reduction for the Very Large Array, the most productive ground-based telescope in the world. It is the primary reduction system for most Very Long Baseline Interferometry including the VLBA and has been used to reduce data from other radio interferometers and single-dish telescopes as well as data taken at other wavelengths. The history and general structure of this software package are reviewed and a number of the scientific achievements for which it has been used are summarized.

1. Early history

The National Radio Astronomy Observatory was established in 1956 for the purpose of building large radio telescopes and operating them as a service to the world's astronomers. The first telescopes built, a 300-foot diameter transit telescope and an 85-foot and a 140-foot fully steerable telescope, were very productive. However, particularly with the receiver technologies of the 1960s and 1970s, these tended to produce data at relatively low rates. Such data required a modest software effort at most and much of the programming was left to the visiting and staff scientists.

Even in the 1950s however, it was realized that a radio interferometer was needed in order to achieve high sensitivity and spatial resolution. Preliminary design work was begun on the Very Large Array as early as 1960 (Heeschen 2000). Proposals for the VLA were issued in 1967 and 1969 and

[†]The National Radio Astronomy Observatory is a facility of the (US) National Science Foundation operated under cooperative agreement by Associated Universities, Inc.

A. Heck (ed.), Information Handling in Astronomy – Historical Vistas, 109-125.

first funding for its construction was received in 1972. Construction was completed in 1981 (Thompson *et al.* 1981).

In order to acquire experience with interferometric receivers, correlators, observing, and analysis, a second 85-foot telescope, this time movable, was constructed and first fringes were obtained in 1964. By 1965, serious scientific results from the Green Bank Interferometer were being published; see Hjellming (2000) for a review. Software for the new interferometer was originally written by the scientific staff, and a rather significant package of programs was developed. Eventually, maintenance and development of this package was assumed by the Computer Division.

The software needs for the proposed VLA were underestimated by enormous factors in the proposal stage and continued to be severely misjudged throughout the construction phase, despite the programmers' name for the VLA ("Data Inundation Device"). Numerous committees met and, each time, the number of super computers required for VLA data calibration and analysis went up. Finally, early in 1979, it was decided to begin a software effort in Charlottesville, Virginia, the site of the NRAO's headquarters. It was felt that such a software project would be more immune to the day-to-day needs of the telescope which would, of necessity, plague the parallel software efforts being conducted at the VLA site in New Mexico.

The design discussions for this project, which had been going on for some time, had come up with a list of basic criteria for the new system. The first of these criteria was that the software had to be transportable across both space and time. Although it was assumed that observers would come to NRAO to do the main reductions of their data, many of them would have obligations at their home institutions that would prevent them from completing their analysis during their visits. Furthermore, it was very clear that the NRAO could not afford the computing resources that the instrument demands. If a software package may be run at the user's home institution, then the usage time and computer power available to the astronomer are increased significantly.

The software needs to be general and flexible to allow for changes in instrumentation at the VLA and for use on data from other telescopes. The software must be interactive in order to allow the user to determine the progress of the reductions and to discover, and point out to the software, interesting regions within the data. Given the large volumes of data, the software must be efficient and make as good use of the computer's capabilities as possible. It should be "friendly" – relatively easy for beginning users and yet powerful for advanced users. There should be easy access to the system both for multiple interactive users and for users with long, batch-like computations. In order to survive over time and changes in personnel, the system must be well documented and there must be uniformity in coding standards, conventions, service routines, and the like. These criteria form the basis of the system we now know as AIPS (Astronomical Image Processing System).

I have written two review articles, one covering the history of AIPS' first 10 years in some detail (Greisen 2000) and one for a summer school using AIPS as a tutorial example of system design (Greisen 1988). The software effort began in the middle of 1979 using a ModComp computer equipped with a Floating Point Systems AP120B array processor and an International Imaging Systems M70 television display device. This computer used 16-bit word addressing, which forced all programs to be smaller than 128 kilobytes (after much overlaying). In fact, the oversight committees wanted us to make the programs half that size so that they would run on Digital Equipment Corporations PDP 11s. We never made it that small. At the end of 1979, we acquired a DEC VAX 11/780 with an AP and TV, which gave us our first taste of a modern software development environment.

When we began the project, "portable" software was a package of a few thousand lines of Fortran that could be brought up on another computer system in under a year. With AIPS, we achieved immensely more than that, but at the cost of some serious complexities in the software. Fortran 77 was not widely available and PDP 11's had hardware unable to meet the requirements of that language. Therefore, we used Fortran 66, an ill-defined language with numerous portability issues which were solved but required significant effort from the programmer. This early form of AIPS eventually ran on machines ranging from the ModComp and VAXes with and without array processors to UNIX desktop computers, IBM mainframes, and even Cray supercomputers.

AIPS had to contend with political matters. In its earliest days, there was an oversight committee made up primarily of scientists actually stationed at the VLA site. They were understandably nervous about a system being built 2000 miles away by a group over which they had little control. By renaming the project RANCID (Radio Astronomy Numerical Computation and Imaging Device) for a year, we managed to get people to take us less seriously long enough for us to get the project well underway. When the Director had to explain to his oversight committees the expenditure of one million (1979) dollars on this project, we renamed it AIPS, which was at least a little more serious.

Quite a number of people have made significant contributions to AIPS over the years. Of these, Bill Cotton is by far the most important. His early work on high-performance imaging and later design and coding of the calibration and object-oriented software make AIPS what it is today. David Brown, Phil Diamond, Chris Flatters, Walter Jaffe, Pat Murphy, and

Don Wells should also be mentioned here, while a more complete cast of characters may be found in Greisen (2000). At its peak, development and support of AIPS reached about 7 full-time employees for a short period; 3–4 was more typical, and the current number is between 1 and 2 for the combination of development and user support.

2. Structure of AIPS

AIPS was begun at about the same time as FITS was designed – see Greisen (2002) in this volume – and was the first system profoundly affected by FITS concepts. The internal header format of AIPS is a binary representation of a FITS header and the disk structure consists of a header file, an image or uv-data file, and "extension" files which are mostly tables, plot files, and history files. It is now written in ANSI-standard Fortran 77 with some system-dependent and X-Windows routines written in ANSI-standard C. Certain system functions are performed by scripts (now mostly Bourne-shell UNIX scripts) and recently some have been recoded and extended using Perl, particularly the installation script.

AIPS consists of a main program with which the user interacts, also called `AIPS`, plus a large set of separate programs which are started by the main program and which may also interact with the user. The structure is illustrated in Fig. 1. `AIPS` takes input interactively from the user or from a text "RUN" file. The input is interpreted by a simple, but essentially complete, computer language called POPS which allows the usual mathematical, logical, and string operations and the creation and execution of procedures which may call other procedures. Some application functions, such as interactive display functions, are run within the main program. Heavily computational operations are run synchronously or asynchronously in separate programs called tasks.

These tasks may be developed and run with no modification to the main program. The only interaction between `AIPS` and the task is through a well-defined interface of parameters, called adverbs, whose values are set by the user in `AIPS` and then sent to the task at start-up time[1] and, of course, through any changes that the task may make to the data on disk. Tasks may also be interactive, in which case they run synchronously and retain control of the user input terminal (now usually an xterm window) and other interactive devices such as the TV display. The TV and graphics displays are now done with programs that emulate the earlier devices. XAS emulates the full IIS M70 on X-Windows screens that are capable of full TrueColor display and implements the image catalog as well, while line graphics is done with a Tektronix emulation found in standard xterm and

[1]The option to allow tasks to return adverb values to `AIPS` was finally added in 2002.

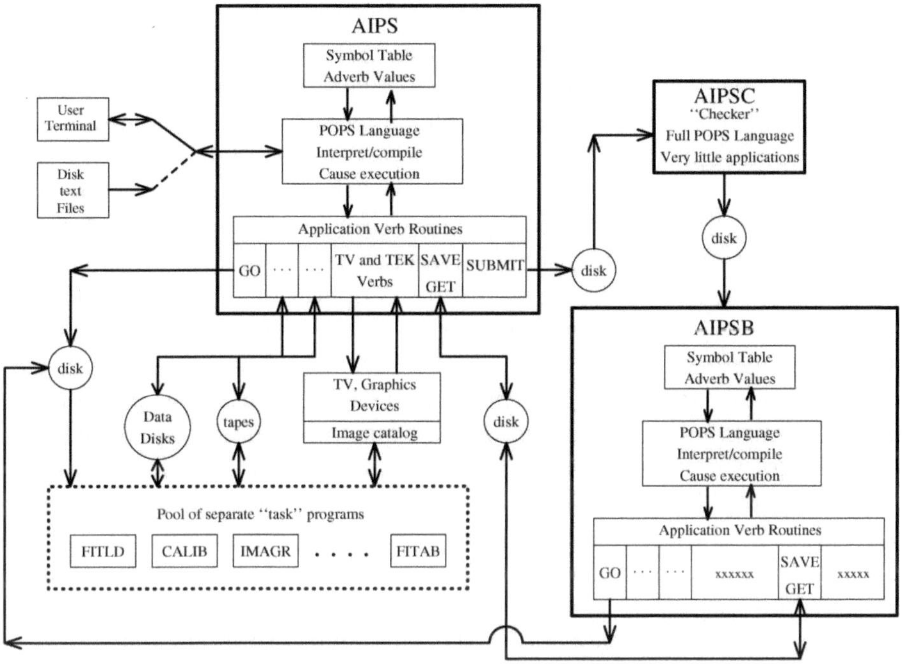

Figure 1. Block diagram of AIPS from a user point of view. Various communications paths are shown among the main interactive program, AIPS, the batch program AIPSB, and the collection of separate tasks.

other terminal programs. The ability to run a full batch queue with a batch-control version of AIPS, called AIPSB, was used heavily in the early days and still remains available.

An image processing system must perform operating-system-like functions. It must, for example, be able to create, open, write, close, truncate, reopen, read, and delete disk files. It must know the current time and date, talk to display devices, enable multiple processes to communicate, and translate between internal and external binary data representations. In order to do this, while keeping the majority of the code portable, we invented[2] the concept of a "virtual operating system" or, more generally, "virtual device interface." In AIPS, we designed what we thought an operating system should look like and then forced each system to behave that way with a layered set of "Z" routines. The layers of Z routines are structured to achieve maximal reuse of code, so that routines that work for most

[2]Others also invented these concepts independently and much study in computer science was devoted to them, all unknown to us.

systems are at the top and then branches separate, for example, VMS and UNIX, then separate flavors of UNIX, and only then the various vendors.

The other virtual device interfaces in AIPS are for array processors, interactive (TV) displays, and graphics displays. In the early days, at great effort, we achieved barely tolerable performance on intensive computational tasks by writing special code for our array processors. That code was separated out as "Q" routines and an emulation of the array processor was coded to use the host CPU directly. Today, array processors cost too much to code and all applications use the pseudo-array processor software. Because these subroutines isolate intensive computational operations, compilers are often able to create very efficient executables from them. In the early days, there were a large variety of TV-like image displays. By designing an interface to an ideal display (actually much like our initial IIS M70), we were able to code "Y" routines for a large number of routines including the ones for `XAS`, the X-Windows emulation of the IIS M70, used by all current AIPS installations. The graphics functions are isolated by having all plot programs call a limited set of graphics routines which write their plot commands into disk files attached to the primary image or uv-data file. Programs for each possible graphics display (*e.g.,* TV, Tektronix, PostScript printer) were then written to interpret the plot file to the particular output device.

AIPS has been widely distributed, reaching all continents and most states in the U.S. We used to poll known sites annually, but gave it up after 1990. Compiling the poll was an enormous job and desktop workstations made our standard questions (how many users, what fraction of the CPU time for AIPS, etc.) less meaningful; see Bridle and Nance (1991) for details of the last formal poll. These polls and our current informal registration system have tracked the move from VAX VMS to SUN workstations to PCs running Linux and the continued increase in users of, and the compute power used for, the software. Information about AIPS, and the code itself, is available at no charge from `http://www.aoc.nrao.edu/aips`.

3. Later history

Tests in the late 1980's revealed that compilers produced less efficient, and sometimes incorrect, code when forced to use the Fortran 66 dialect of AIPS, rather than a standard Fortran 77. That language was finally available everywhere and offered numerous advantages including true character strings, `IF-THEN-ELSE` constructs, and defined data sizes which greatly simplified the calling of subroutines and the use of data structures. We therefore froze AIPS for nearly 18 months and completely overhauled the code, releasing the new version finally in October 1989. It was in this same period that the VLA decommissioned its old DEC-10 with its familiar VLA calibration software. A large number of users then found themselves dependent on the new, and largely untried, calibration software in AIPS. That package was designed with more general interferometers, particularly very long baseline (VLBI), in mind and, hence, was unfamiliar to VLA observers.

An unfortunate combination then occurred of the discontent over the new calibration package, the long-standing discomfort caused by dependency on software generated elsewhere, and various feuds that had arisen over software issues, particularly AIPS' adherence to strict coding and generality standards. Meetings were then held to determine the future of NRAO's software effort. These meetings concluded that AIPS should be overhauled substantially, one major piece at a time. Nonetheless, the new head of the Computer Division Geoff Croes decided in 1990 to create a new software package using the new `C++` language. I will not attempt to describe that project, called `aips++`, here. It is a major software project, involving an international consortium, and a significant effort in manpower, and is aimed at various aspects of telescope control and data handling in addition to basic synthesis reduction. While `aips++` is not quite ready for production use in reducing VLA and VLBA data, it has already created very useful libraries, special-purpose production packages, and interesting new algorithms. More information is available at `http://aips2.nrao.edu/`.

Croes wrote in the April 15, 1991 AIPSLetter: "A major consequence of this plan is that AIPS has been frozen. Current software development in AIPS is limited to improvements needed in the short term for VLBA calibration and certain experimental imaging tasks ... the manpower we have available to support AIPS has been dramatically reduced ..."

Despite this edict, there have been significant developments in AIPS in the last 11 years. AIPS was converted to run across Local Area Networks, freely sharing disks, tape drives, and displays between computers. The X-Windows display program was overhauled to support TrueColor and to be self-describing. Bill Cotton wrote a large package of data interface routines based on the concepts of object-oriented methodology (Cotton 1992). This package made possible the construction of tasks much larger than those which were maintainable under traditional software methodologies. `IMAGR`, to be described in the next section, is a major example. In addition, significant, nearly revolutionary, improvements have been made in all phases of the package from data loading, editing, and calibration to imaging, display, and data storage. However, because of its putative limited lifetime and lack of manpower, the AIPS group has not carried out major overhauls of whole areas of code/design. It has also become increasingly difficult to attract and retain employees in the group.

4. Algorithmic achievements

The main role of a package like AIPS is not to experiment with new algorithms. Instead, it is to deliver a reliable and comprehensive package of algorithms in a form that can be used by a wide range of scientists around the world. This, AIPS has done. Nonetheless, users have been drawn to the package on several occasions by algorithms that had their first widely available implementation in AIPS.

Images produced by an interferometer are convolved with the Fourier transform of the data sampling, which is a quite complicated function. Högbom (1974) invented an iterative algorithm called Clean for deconvolving such images and Clark (1980) not only invented an efficient implementation of Clean, but also made his code for the array processor available to the AIPS group. This was the only widely available implementation of this particularly efficient algorithm. Cotton and Fred Schwab devised an enhancement of this algorithm which periodically subtracts the Clean component model from the visibility data and makes new residual images. Since this allows images to be Cleaned almost to their edge and permits numerous corrections to be made in the process, the Cotton/Schwab Clean is the one used by `IMAGR`.

Images produced by an interferometer are also affected by errors in the gain and phase calibration, often due to short-term changes in the atmosphere or ionosphere. If the model of the source were perfect, then any difference between the observed and model visibilities must be due to calibration error. Since the calibration generally is assumed to be antenna dependent, there are only N gains to be determined from the observations made on $N(N-1)/2$ antenna pairs. If the number of antennas is significant, this is a heavily over-determined problem which means that errors in the source model will not overly degrade the gains derived. This "self-calibration" may be iterated with re-imaging of the corrected data until the solutions converge. This algorithm was first implemented by Schwab (1980) in AIPS.

The effects of Clean and self-calibration on a VLA image are illustrated in Fig. 2, which shows a radio galaxy observed by Clarke & Vogt (2002). Without Clean and self-calibration, the jets of this (and most) galaxies would not be visible.

The task `IMAGR` first appeared in 1995. In its initial release, it implemented numerous options for data weighting including the "robustness" concept (Briggs 1995) which allowed an optimal balance of spatial resolution and signal-to-noise to be achieved. Images produced by an interferometer are correct on the celestial sphere, not the two-dimensional planes on which they are usually computed. Beginning in 1998, `IMAGR` is able to do the ima-

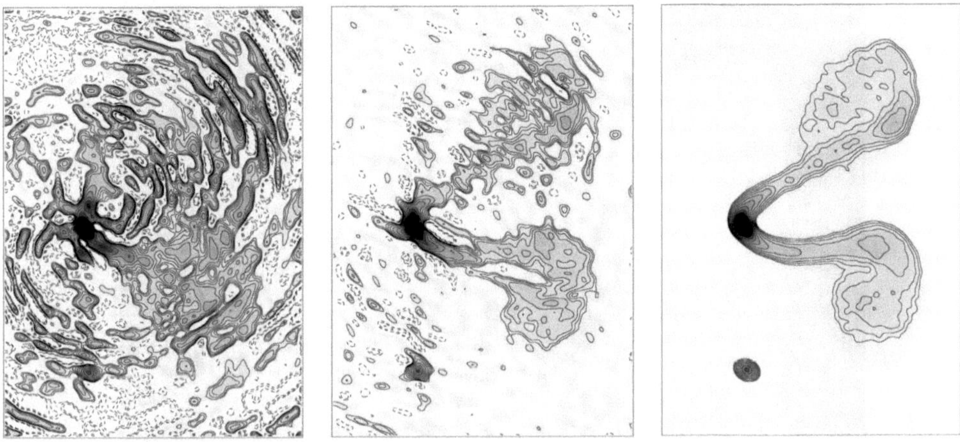

Figure 2. Clarke & Vogt (2002) images of the radio source IC 708 in the cluster of galaxies known as Abell 1314. The "dirty" image at left has a peak-to-rms ratio of 67. For the Cleaned image in the center that ratio is 262, while it is 2061 for the self-calibrated and Cleaned image at right. The same contours are drawn in each image.

ging on numerous small two-dimensional planes covering the desired field of view and each tangent to the celestial sphere. Although this means of correcting the "3D" problem was described by Cornwell & Perley (1992), it became available to a large number of users only with its implementation in AIPS. Without this option, the very wide-field imaging now done with the VLA at wavelengths $\geq$ 21 cm would not be possible. `IMAGR` now also implements the multi-scale algorithm described by Holdaway & Cornwell (1999). See Fig. 3 for an example. Since AIPS is a complete system, images made with these options produce source models which may be used throughout the package for such operations as self-calibration.

Experiments with maximum entropy related deconvolution techniques by Cornwell & Evans (1985) led Cornwell to place the task `VTESS` in AIPS. This task is still used as a benchmark by which other multiple-pointing ("mosaicing") deconvolution algorithms are measured.

In 1981, NRAO began design studies for an array of telescopes to be spread across the United States and operated as a single very-long-baseline telescope. Initial funding for VLBA construction was received in 1984 and construction was completed in 1993 (Napier *et al.* 1994). On May 6, 1992, the VLBA Correlator produced its first fringes. Within 12 hours and 2000 miles away, AIPS read, sorted, and plotted these data with correct annotation. AIPS has been the main software system of the VLBA and one of the primary systems for all VLBI reductions. In part this is due to an algorithm designed by Schwab & Cotton (1983) to use data from all baselines

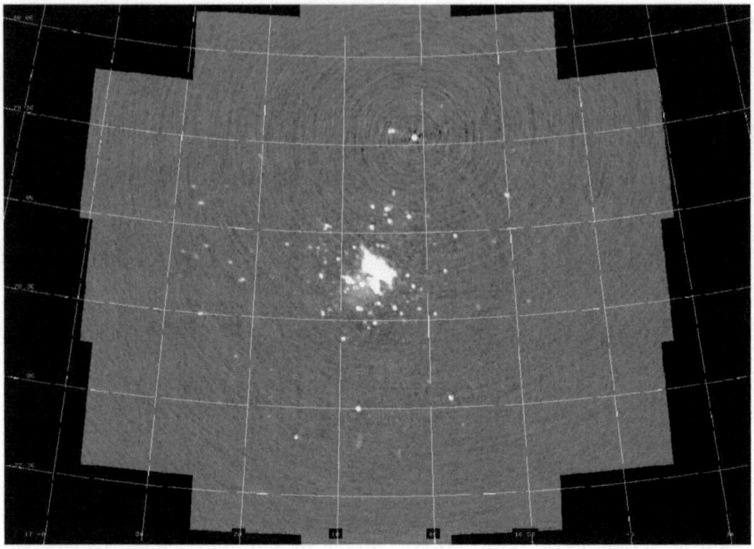

Figure 3. Clarke & Ensslin (2001) data on galaxy cluster Abell 2256 from the VLA's most compact configuration. The image was constructed with 19 facets each tangent to the sphere and then re-sampled into a single plane for this display. The primary beam of the individual antennas in the interferometer covers only a little more than one of the coordinate "squares," but all 19 must be Cleaned to remove sidelobes of objects to which the VLA responds.

simultaneously to make the array maximally coherent, finding corrections on an antenna basis rather than on a baseline basis. Calibrations of greater reliability and accuracy are determined in this fashion and may be found for much weaker sources than previously.

5. Scientific results

According to publication statistics maintained by the NRAO, the VLA has been the most productive ground-based telescope, lagging only the Hubble Space Telescope and IRAS in overall publication rate. (See also Benn & Sanchez (2001) for publication statistics heavily weighted by citations.) The VLA has been used for projects touching all areas of astronomy and AIPS was used for part or all of the data reductions in almost every case. It is not possible in this limited space to even list the major achievements of the instrument; the projects described here are just a nearly random sample influenced by ready availability. The NRAO is developing an image gallery at `http://www.nrao.edu/imagegallery` which includes a wide range of images produced by NRAO's telescopes. The book *Radio Interferometry: The Saga and the Science* published by NRAO has review papers covering many of the important research areas.

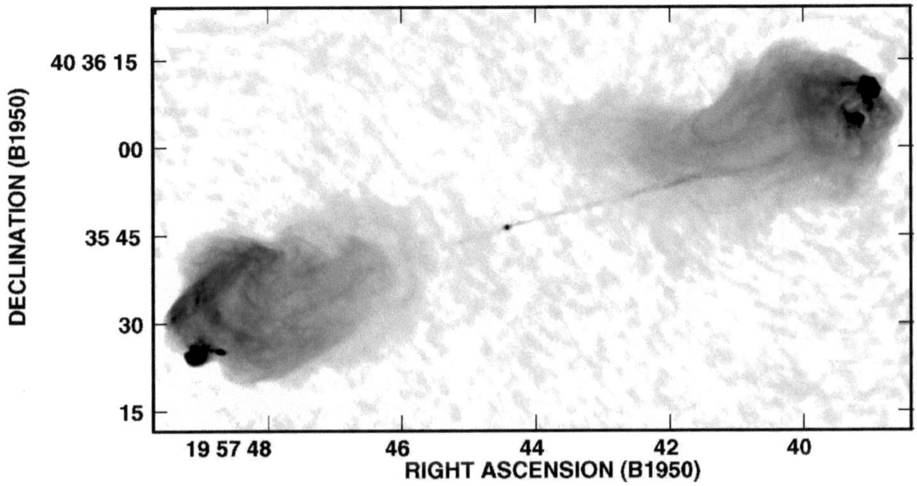

Figure 4. Perley *et al.* (1984) image of the radio galaxy Cygnus A at 4.525 GHz with 0.4 arc second resolution.

The VLA first made a splash with images of radio galaxies such as that shown in Fig. 4. The high angular resolution and significant dynamic range achieved allowed the tubes of plasma, by which energy and matter are transmitted from the central engine to the radio lobes (the "jets"), to be studied for the first time. Images made in the linearly polarized emission permitted the structure of the magnetic fields to be determined. Details in the structure of the radio lobes, suggesting how they are formed, were also revealed; see Owen *et al.* (2000) for an example. The above-mentioned web site has links to Alan Bridle's marvelous collection of images and information on radio galaxies; see also Bridle (2000). The VLBA, with its very high spatial resolution, may be used to study the inner workings of these radio galaxies and quasars (*e.g.*, Zensus *et al.* 2002; Zensus 1997). One of the more important results from the VLBA is a direct measurement of the distance to the active galaxy NGC4258 from measurements of the orbital motions of gas around its nucleus (Herrnstein *et al.* 1999). These data provide a direct measurement of the mass of the central object in that galaxy and are the most direct proof yet of the existence of black holes.

The VLA has also been used for observations of the interstellar medium, particularly the neutral hydrogen emission, of normal galaxies. Fig. 5 shows an example of the total hydrogen content and velocity field of the galaxy NGC6503. Van Gorkom (2000) provides a review, while the NRAO image gallery has a large number of images from this area of research.

The VLA and VLBA have also been used extensively for observations

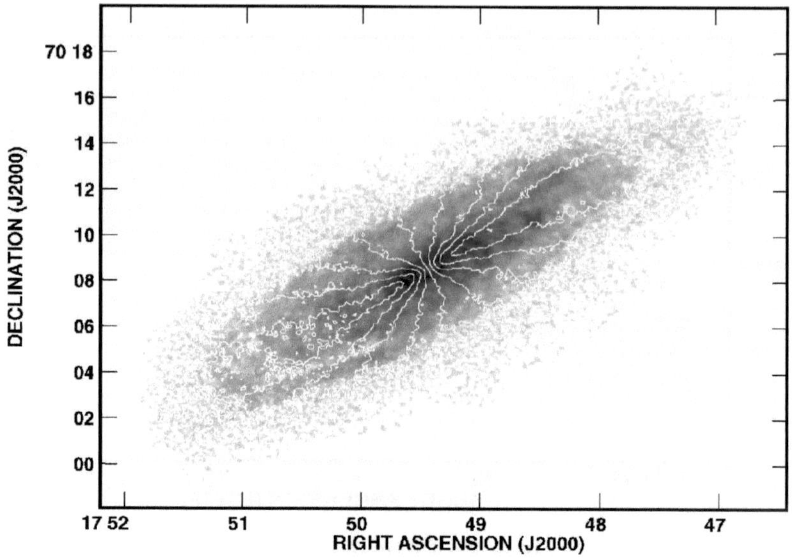

Figure 5. Greisen & van Moorsel unpublished image of the total neutral hydrogen (gray scales) in the galaxy NGC6503 with contours of constant velocity (separated by 25 km/sec) made with the VLA at 14 arc second resolution.

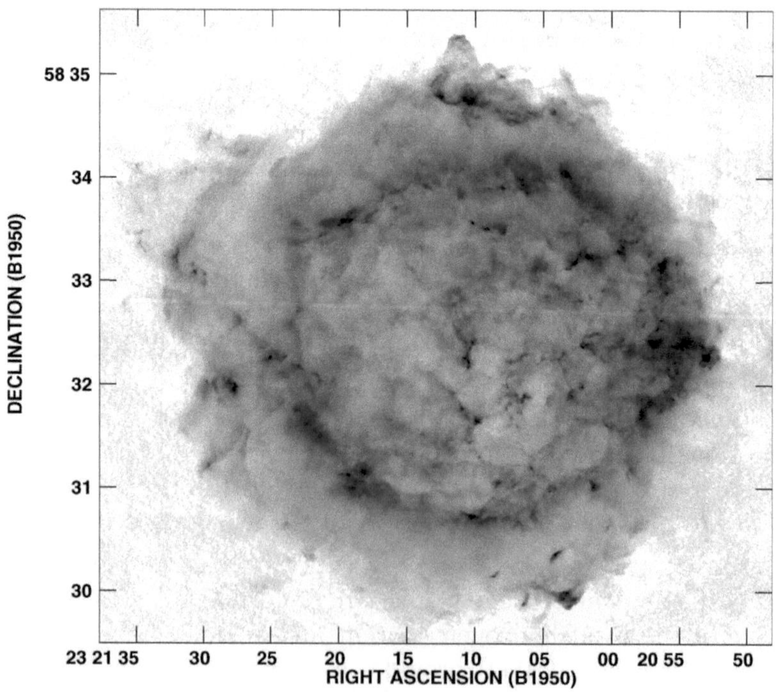

Figure 6. The supernova remnant Cassiopeia A at 1.38 GHz (Anderson *et al.* 1991).

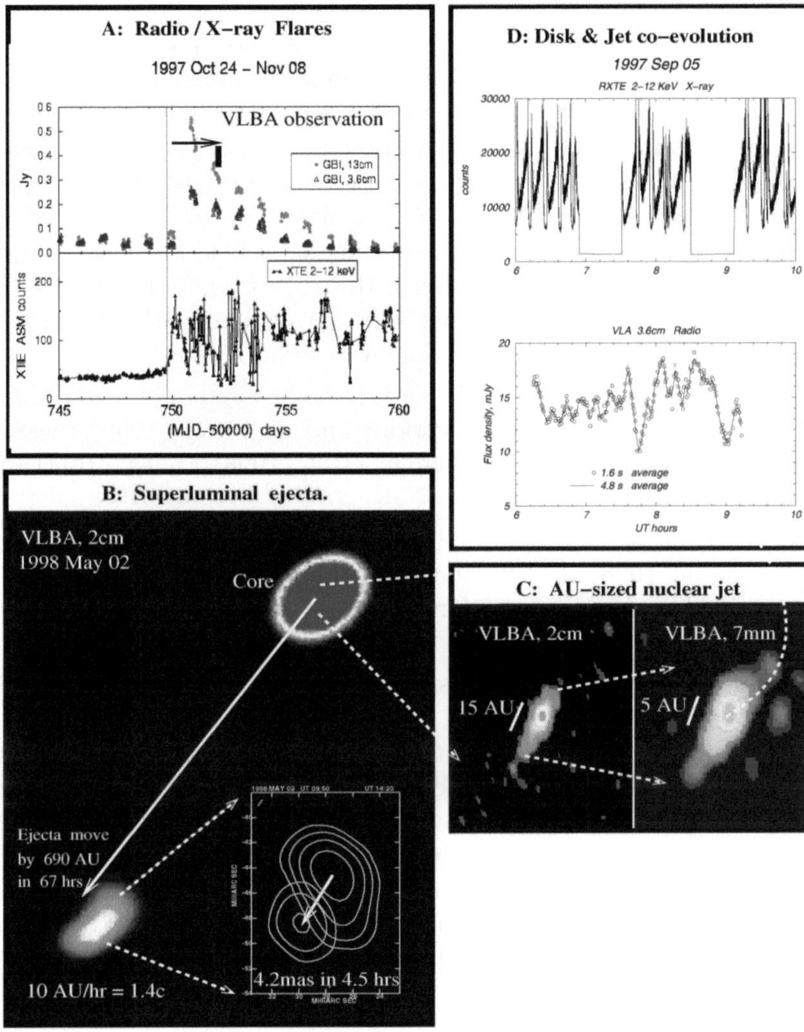

Figure 7. Dhawan *et al.* (2000) images of the rapidly varying Galactic micro-quasar GRS 1915+105.

of objects within our Galaxy. An early image that strained our computing resources is shown in Fig. 6. Supernova remnants have many moving knots of small angular scale, but the bulk of the emission comes from a relatively smooth region of significant angular size which is difficult for interferometers to image. The Galactic Center is another region which is filled with both large-diameter sources as well as incredible narrow arcs and other emission features which appear to trace magnetic fields. The optical light from the center is completely obscured, making radio wavelengths the main source of information about the heart of our galaxy. See Morris (2000) for a review of the VLA's contributions in this area. Most normal stars are relatively weak radio sources, but the VLA has made important contributions to their study (White 2000). Most small-size galactic X-ray sources arise in binary systems in which one star dumps its material onto a compact star, presumably a black hole, neutron star, or white dwarf. These objects can be highly variable, changing on time scales of seconds to minutes while ejecting blobs of charged particles and magnetic field at *apparent* speeds in excess of the speed of light (Mirabel & Rodriguez 1999). GRS1915+105 has been observed nearly simultaneously in X-rays and with the VLBA. The results are summarized in Fig. 7 (Dhawan *et al.* 2000).

The VLA has also been used to study Solar System objects. Muhleman *et al.* (1995) review radar observations and de Pater (1990) reviews synthesis observations of planets. Bastian *et al.* (1998) review observations of the Sun in its active phases.

The two major surveys done by the VLA, the low resolution NVSS (Condon *et al.* 1998) and higher resolution FIRST (Becker *et al.* 1995), have provided source material for many statistical studies of radio sources of various kinds.

AIPS has also been used for single-dish data when the data rates have gotten large. This first arose with all-sky surveys conducted with the Green Bank 91-m (300-foot) telescope at 1.4 GHz (Condon & Broderick 1985 and 1986) and then at 4.85 GHz (Condon *et al.* 1994). AIPS was then used with millimeter-wavelength data taken by the 12-m telescope in Tucson in the popular on-the-fly observing mode. This technique is described by Mangum *et al.* (2000) and scientific results obtained with the 12-m using AIPS include Womack *et al.* (2000), Arce & Goodman (2000), and Koo (1999). The 12-m was even used in beam-switched on-the-fly continuum mode with the data reduced in AIPS by Liszt & Lucas (1999).

6. Summary

AIPS, the Astronomical Image Processing System, has served the NRAO for 23 years. Its main role has been to put interferometric data calibration, editing, imaging, and display programs in the hands of the users of NRAO's Very Large Array and Very Long Baseline Array, both at the NRAO and at their home institutions. It has also provided useful functionality for data from other radio interferometers, single-dish radio telescopes, and even optical telescopes. With these tools, the users have produced a dazzling array of important and useful scientific results from nearby comets, planets, and the Sun to strange and wonderful objects within our galaxy to powerful radio galaxies and quasars at the very edges of the universe.

Acknowledgements

The author thanks Michael Rupen and Jim Ulvestad for assistance with this manuscript and Tracy Clarke, Vivek Dhawan, Gustaaf van Moorsel, and Frazer Owen for providing image data.

References

1. Anderson, M., Rudnick, L., Leppik, P., Perley, R.A. & Braun, R. 1991, Relativistic electron populations in Cassiopeia A, *Ap. J.* **373**, 146.
2. Arce, H.G. & Goodman, A.A. 2000, On-the-fly mapping of giant molecular outflows, *Imaging at Radio Through Submillimeter Wavelengths*, Mangum, J. G., Radford, S. J. E Eds., *A. S. P. Conference Series* **217**, 86.
3. Bastian, T.S., Benz, A.O. & Gary, D.E. 1998, Radio emission from solar flares, *Ann. Rev. of Astr. & Ap.* **36**, 131.
4. Becker, R.H., White, R.L. & Helfand, D.J. 1995, The FIRST survey: faint images of the radio sky at twenty centimeters, *Ap. J.* **450**, 559. See also `http://sundog.stsci.edu/top.html`.
5. Benn, C.R. & Sanchez, S.F. 2001, Scientific Imapct of Large Telescopes, *P. A. S. P.* **113**, 385.
6. Bridle, A.H. 2000, Impact of the VLA: physics of AGN jets, *Radio Interferometry, The Saga and the Science*, National Radio Astronomy Observatory Workshop **27**, Socorro, 152.
7. Bridle, A. & Nance, J. 1991, The 1990 AIPS site survey, *AIPS Memo Series* **No. 70**, National Radio Astronomy Observatory, Charlottesville.
8. Briggs, D.S. 1995, High fidelity deconvolution of moderately resolved sources, Ph.D. thesis, The New Mexico Institute of Mining and Technology, Socorro.
9. Clark, B.G. 1980, An efficient implementation of the algorithm "Clean," *Astr. & Ap.* **89**, 377.
10. Clarke, T.E. & Ensslin, T.A. 2001, Cluster Mergers and Diffuse Radio Emission in Abell 2256 and Abell 754, to appear in *Galaxy Clusters and the High Redshift Universe Observed in X-rays*, Neumann, D., Durret, F., and Tran Thanh Van, J. Eds., Proceedings of XXI Moriond Astrophysics Meeting, in press, astro-ph/0106137.
11. Clarke, T.C. & Vogt, C. 2002, in preparation.
12. Condon, J.J. & Broderick, J.J. 1985, A 1400 MHz sky survey. I. confusion limited maps covering $7^h30^m < \alpha < 19^h30^m, -5° < \delta < +82°$, *A. J.* **90**, 2540.
13. Condon, J.J. & Broderick, J.J. 1986, A 1400 MHz sky survey. II. confusion limited maps covering $19^h30^m < \alpha < 7^h30^m, -5° < \delta < +82°$, *A. J.* **91**, 1051.
14. Condon, J.J., Broderick, J.J., Seielstad, G.A., Douglas, K. & Gregory, P.C. 1994, A 4.86 GHz sky survey. III. epoch 1986 and combined (1986+1987) maps covering $0° < \delta < 75°$, *A. J.* **107**, 1829.
15. Condon, J.J., Cotton, W.D., Greisen, E.W., Yin, Q.F., Perley, R.A., Taylor, G.B. & Broderick, J.J. 1998, The NRAO VLA Sky Survey, *A.J.* **115**, 1693. See also `http://www.cv.nrao.edu/nvss/`.
16. Cornwell, T.J. & Evans, K.F. 1985, A simple maximum entropy deconvolution algorithm, *Astr. & Ap.* **143**, 77.
17. Cornwell, T.J. & Perley, R.A. 1992, Radio-interferometric imaging of very large fields the problem of non-coplanar arrays, *Astr. & Ap.* **261**, 353.
18. Cotton, W.D. 1992, Object oriented programming in AIPS fortran (OOPS), *AIPS Memo Series* **No. 78**, National Radio Astronomy Observatory, Charlottesville.
19. de Pater, I. 1990, Radio images of the planets, *Ann. Rev. of Astr. & Ap.* **28**, 347.
20. Dhawan, V., Mirabel, I.F. & Rodriguez, L.F. 2000, AU-scale synchrotron jets and superluminal ejecta in GRS 1915+105, *Ap. J.* **543**, 373.

21. Greisen, E.W. 1988, The Astronomical Image Processing System, *Acquisition, Processing, and Archiving of Astronomical Images*, Longo, G., Sedmak, G. Eds., Osservatorio Astronomico di Capodimonte & FORMEZ, Naples, 125.
22. Greisen, E.W. 2000, The VLA—AIPS, *Radio Interferometry, The Saga and the Science*, National Radio Astronomy Observatory Workshop **27**, Socorro, 57.
23. Greisen, E.W. 2002, FITS: a remarkable achievement in information exchange, *Information Handling in Astronomy – Historical Vistas*, Heck, A. Ed., Kluwer Academic Publishers, Dordrecht (this volume).
24. Heeschen, D.S. 2000, The VLA – planning and construction, *Radio Interferometry, The Saga and the Science*, National Radio Astronomy Observatory Workshop **27**, Socorro, 46.
25. Herrnstein, J.R., Moran, J.M., Greenhill, L.J., Diamond, P.J., Inoue, M., Nakai, N., Miyoshi, M., Henkel, C. & Riess, A. 1999, A geometric distance to the galaxy NGC4258 from orbital motions in a nuclear gas disk, *Nature* **400**, 539.
26. Hjellming, R.M. 2000, Science with the Green Bank Interferometer, *Radio Interferometry, The Saga and the Science*, National Radio Astronomy Observatory Workshop **27**, Socorro, 28.
27. Högbom, J.A. 1974, Aperture synthesis with a non-regular distribution of interferometer baselines, *Astr. & Ap. Suppl. Series* **15**, 417.
28. Holdaway, M. & Cornwell, T. 1999, Multi-scale clean, Mosaicing Workshop, Socorro, NM, see `http://aips2.nrao.edu/docs/user/General/node16.html`
29. Liszt, H. & Lucas, R. 1999, 86 and 140 GHz radiocontinuum maps of the Cassiopeia A SNR, *Astr. & Ap.* **347**, 258.
30. Koo, B.-C. 1999, CO observations of the W51B HII region complex, *Ap. J.* **518**, 760.
31. Mangum, J.G., Emerson, D.T. & Greisen, E.W. 2000, The on-the-fli imaging technique, *Imaging at Radio Through Submillimeter Wavelengths*, Mangum, J. G., Radford, S. J. E Eds., *A. S. P. Conference Series* **217**, 179.
32. Mirabel, I.F. & Rodriguez, L.F. 1999, Sources of relativistic jets in the galaxy, *Ann. Rev. of Astr. & Ap.* **37**, 409.
33. Morris, M. 2000, The VLA in galactic center research, *Radio Interferometry, The Saga and the Science*, National Radio Astronomy Observatory Workshop **27**, Socorro, 112.
34. Muhleman, D.O., Grossman, A.W. & Butler, B.J. 1995, Radar investigations of Mars, Mercury, and Titan, *Ann. Rev. of Earth and Planetary Sciences* **23**, 337.
35. Napier, P.J., Bagri, D.S., Clark, B.G., Rogers, A.E.E., Romney, J.D., Thompson, A.R. & Walker, R.C. 1994, The Very Long Baseline Array, *Proc. IEEE* **82**, 658.
36. Owen, F.N., Eilek, J.A. & Kassim, N.E. 2000, M87 at 90 centimeters: a different picture, *Ap. J.* **543**, 611.
37. Perley, R.A., Dreher, J.W. & Cowan, J.J. 1984, The jet and filaments in Cygnus A, *Ap. J.* **285**, L35.
38. Schwab, F.R. 1980, Adaptive Calibration of Radio Interferometer Data, *Proc. Soc. Photo-Opt. Instrum. Eng.* **231**, 18.
39. Schwab, F.R. & Cotton, W.D. 1983, Global fringe search techniques for VLBI, *A. J.* **88**, 688.
40. Thompson, A.R., Clark, B.G., Wade, C.M. & Napier, P.J. 1980, The Very Large Array, *Ap. J. Suppl.* **44**, 151.
41. van Gorkom, J. 2000, HI images that change my view of the universe, *Radio Interferometry, The Saga and the Science*, National Radio Astronomy Observatory Workshop **27**, Socorro, 137.
42. White, S.M. 2000, The contributions of the VLA to the study of radio stars, *Radio Interferometry, The Saga and the Science*, National Radio Astronomy Observatory Workshop **27**, Socorro, 86.
43. Womack, M., Pinnick, D.A., Mangum, J.G., Festou, M.C. & Stern, S.A. 2000, On-the-fly imaging of neutral and ionized molecules in Comet Hale-Bopp, *Imaging at*

Radio Through Submillimeter Wavelengths, Mangum, J. G., Radford, S. J. E Eds., *A. S. P. Conference Series* **217**, 82.

44. Zensus, J.A. 1997, Parsec-scale jets in extragalactic radio sources, *Ann. Rev. of Astr. & Ap.* **35**, 607.
45. Zensus, J.A., Ros, E., Kellerman, K.I., Cohen, M.H., Vermeulen, R.C. & Kadler, M. 2002, Sub-milliarcsecond imaging of quasars and active galactic nuclei. II. additional sources, *A. J.* **124**, 662.

CHANGES IN ASTRONOMICAL PUBLICATIONS DURING THE 20TH CENTURY

H.A. ABT
Kitt Peak National Observatory
P.O. Box 26732
Tucson, AZ 85726-6732, USA
abt@noao.edu

Abstract. Among the major changes in astronomical publication during the 20th century are the virtual demise of observatory publication, the growth of conference proceedings, and the continuing dominance of journal publications. The numbers of research papers were found to depend only on the number of researchers and not upon the speed of new detectors and computers or the availability of large telescopes. Papers have grown in average length by a factor of 5 but their lengths have leveled off because many data are published on-line only. The fraction of papers with authors from two or more countries is currently 40% and growing by 1% per year. After trying various publication methods (microfiche, CD-ROMs, videos), the trend is toward on-line publication. With the growth and complexity of science, it is increasingly important to obtain independent reviews of papers. Current auxiliary tools include search engines, the Science Citation Index, and preprint servers.

1. Introduction

We review the changes during the 20th century in the types of publication employed, the numbers of papers published, the methods of publication, the characteristics of those papers, and the uses made of them. Although the evidence is taken from the field of astronomy, much of it applies to other sciences.

One major change in publications is difficult to quantify and that is the style of writing. Only by reading papers published early in the century does one appreciate how much the current papers have become filled with

A. Heck (ed.), Information Handling in Astronomy – Historical Vistas, 127-137.

information. For instance, consider the lead paper in the third issue of the *Astronomical Journal* in 1901. The complete text is

> "The following observations of *Eros* were made with the Bruce Micrometer on the twenty-inch equatorial. The magnifying power was two hundred. The right-ascension observations are chronographic, and the declination-bisections were made while the object was drifting through the field."

There follows a four-page table of positions. Notice that there is no introduction that would explain why it is important to measure positions of that asteroid, no discussion of the measurements, and no conclusions. Such an uninformative set of measurements would not have been acceptable for publication later in the 20th century.

Another characteristic of papers early in the century was their limited breadth. All of the papers published in the *Astronomical Journal* during the first few years of the century concerned positional measurements of solar system objects and stars, celestial mechanics, and observations of variable stars. The complimentary journal of the time, the *Astrophysical Journal*, had papers of a broader content, mostly because it included many papers of laboratory spectroscopic measures, new instruments, astronomical spectroscopic studies of binaries, and solar spectra. It would be unfair for us to criticize their lack of astronomical knowledge and tools at the time, but the leisurely manner of their descriptions are no longer acceptable in our concisely written papers.

2. Locations of Publication

A favorite median for publication earlier in the century was the observatory publications. Each major observatory had its own publications series, which was sent free to other observatories in return for receiving theirs. With time it was realized that this system had two defects: (1) the entire cost of publication and distribution fell onto the distributing observatory; the readers paid nothing, and (2) the scientific reviewing of the papers were of unknown rigor.

The frequencies of references in several journals are listed in Table 1. They are taken mostly from Abt (1995a) plus a study of the 1900 *Astrophysical Journal.*

We see that a steady 80% of the referenced papers are to journals, a decreasing fraction to observatory publications, an increasing fraction to conference proceedings, an increase in reviews, and a decreasing fraction to monographs. The other sources provide negligible fractions.

The growth in the number of conference publications is dramatic and there are good reasons for that. The first is that astronomers worldwide wish

TABLE 1. Percentages of various venues for astronomical publications.

Source	1900	1952	1962	1972	1982	1992	Trend
Journals	84	78	76	76	79	80	Constant
Obs. publ.	4	12	7	4	3	1	Decrease
Private com.	0	2	1	2	1	1	Constant
Preprints	0	0	1	1	3	2	Increase
Theses	0	0	1	2	2	1	Constant
Reviews	0	0	1	1	1	2	Increase
Conf. proc.	0	1	5	7	7	10	Increase
Monographs	12	5	8	6	4	4	Decrease

to meet with others working in the same field to exchange ideas and make plans, and travel has become much easier. In fact, as we shall see in Sect. 5, about one-third of the papers currently being published have authors from two or more different countries. Much of such research is facilitated by e-mail, but opportunities to meet in person are cherished. Second, if astronomers cannot attend a conference of particular interest, they can learn from the conference volume. Third, a pragmatic reason is that many astronomers cannot receive travel funds to go to a meeting unless they present a paper. The objection to conference papers is that many of them are preliminary studies and are not reviewed critically, so astronomers are less likely to believe the claimed results than for refereed research papers.

3. Number of Papers

The plot (see Fig. 1) of the logarithm of the number of pages (normalized to 1000-word contents) in the *Astrophysical Journal* during the 20th century is surprising for two reasons. First, after 30+ years of constant content, the numbers of pages of astrophysics increased abruptly around 1932. What happened at that time to start a large burst of publication? That was in the depth of the American depression and no new telescopes or major equipment were put into use. The explanation appears to be that the progress in atomic and molecular physics and gas dynamics made it possible, for the first time, to derive physical parameters, such as abundances, temperatures, densities, excitation conditions, etc. from astronomical spectra. Previously astronomers were able only to *measure* Doppler shifts of stars and other objects, but the papers after 1932 were filled with physically meaningful models of astronomical objects.

The second surprising aspect of Fig. 1 is the steady growth, following

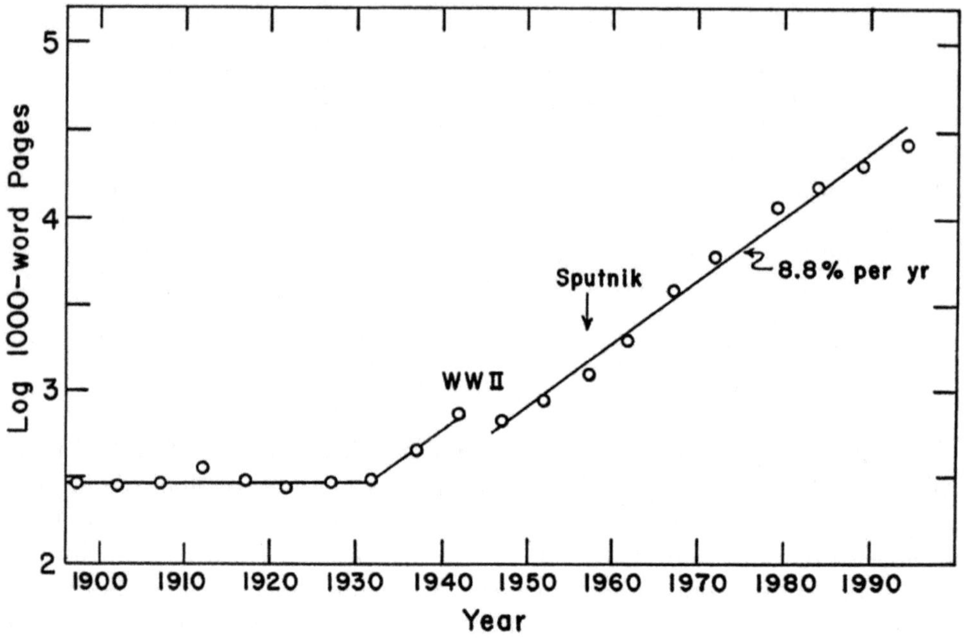

Figure 1. The logarithm of the numbers of pages, normalized to 1000 word contents, published annually in the *Astrophysical Journal* are plotted against time during the 20th century. These data (Abt 1995b) are reproduced by permission from the *Astrophysical Journal.*

the interruption of World War II. There were no significant peaks or bursts with the start of the space age, the increases in speed and availability of computers by many orders of magnitude, the construction of many large telescopes, and the nearly universal replacement of photographic plates with solid-state detectors that are several orders of magnitude faster. Why was that?

The answer was shown in Abt (2000). The growth in the numbers of American astronomers, counted as members of the American Astronomical Society, was parallel to the numbers of papers published in the three major American astronomical journals (ApJ, AJ, PASP). The ratios are shown in Table 2 for the last 30 years.

We see that since 1975 the ratio has been constant at 0.43 ± 0.02 papers per astronomer. Therefore the limitation to our output is the number of astronomers doing research. If they are provided with larger telescopes, faster computers, data from a wide range of wavelengths, etc., they may

TABLE 2. Numbers of papers in ApJ+AJ+PASP, numbers of members of the AAS, and their ratios.

Year	Total Papers	Members of AAS	Papers per Member
1970	908	2619	0.35±0.01
1975	1355	3250	0.42±0.02
1980	1622	3815	0.42±0.02
1985	1723	4284	0.40±0.01
1990	2207	5297	0.42±0.01
1995	2869	6211	0.46±0.02
1999	2944	6745	0.44±0.01

produce better or more useful papers but not more papers. If we want to increase our output of papers, we should employ more astronomers rather than to build more telescopes.

4. Numbers of Authors per Paper

During the first two decades of the century all of the theoretical papers and almost 90% of the observational papers were single authored (Abt 1981). By mid-century only 75% of the papers were single-author ones. A count of papers in the *Astrophysical Journal* in 2000 shows that only 10±2% of such papers are single-authored papers. Although this fraction appears to be asymptotically approaching zero, it seems unlikely that it will ever reach zero.

Similarly the average number of authors per paper has increased from slightly more than 1.0 at the beginning of the century (Abt 1981) to 1.3 at mid-century to 3.8 at the end of the century (Abt 2000). Some fields, such as physics, have some papers with 100 or more authors; in astronomy the large number rarely exceed 25. We could raise the question of the meaning of authorship in such circumstances, but this is not the appropriate place for that discussion.

5. Paper Lengths

Early in the century papers were 2-3 pages (normalized to 1000-word contents) long (Abt 1981). By mid-century they had grown to 5 pages as astronomers put more data, discussion, and results in their papers. Now the average length is asymptotically approached 12-13 pages (Abt 1995b). Part of that leveling off, however, is due to the practice of placing large tables and dozens of illustrations in CD-ROMs or in the on-line editions only. For instance, a recent 8-page paper in the printed edition had 41 pages of tables in the on-line edition. With data flowing in from spacecraft, telescopes with sophisticated equipment, and huge computer simulations and with multiple authors to share the work, the contents of papers is still likely to grow, no matter how concise the writing may be.

6. International Papers

During the first half of the century most of the papers in a national astronomical journal were authored by residents of that country. By 1950 about 10% of the papers came from other countries (Abt 1990). Multiple- authored papers were written almost exclusively by astronomers at the same institution (and same country) or astronomers and their assistants. Papers from astronomers at two different institutions were rare. It was not until about 1970 that 4% of the papers came from authors in two or more countries. That coincided with the advent of e-mail. Now (Abt 2000) about 40% of the papers come from countries other than the country where the journal is published and roughly 30% of the papers are authored by people in two or more different countries. This change has occurred in Europe and Asia, as well as in the US.

Astronomy and geophysics (Abt 1992) both had 26% multi-national papers in 1990; physics, chemistry, and biology had about 12%; mathematics, radiology, and psychiatry about 5%. Therefore the first two are the major international fields in terms of collaborations.

There are several reasons for the current frequent international papers. One is the ease of e-mail for communication. Another is the widespread placement of telescopes in the best observing sites (*e.g.* Chile, Hawaii, Canary Islands) that cause astronomers to travel and encounter astronomers from different countries. Another is the desire to combine observations of different wavelengths or different techniques (*e.g.* imaging, spectroscopic, photometric) that often requires data from several observers who may be in different institutions or countries. Astronomy is still a relatively small field (7000 members of the American Astronomical Society in the USA, 10,000 members of the International Astronomical Union), so astronomers often reach abroad to find collaborators with similar interests.

7. Changing Publication Techniques and Funding

Early in the century the publishers used monotype machines, followed by Linotype machines. In 1976 the *Astrophysical Journal* (ApJ) changed to computerized composition. Each step was less labor-intensive and reduced costs.

With the use of computers the handling of large amounts of data became desirable and possible. In 1974 the ApJ started a microfiche edition to help solve the problem of excessive shelf space for that journal was approaching 1.2 meters of shelf space a year. In 1992 the Journal started a series of video tapes, distributed free to all subscribes, to present time- dependent sequences and numerical simulations. In 1993 it started a series of CD-ROMs to provide data in a form that could be read directly by computers.

Those temporary techniques were replaced in September 1995 with the on- line edition. That edition for the ApJ, and later for the AJ and PASP, was organized by Peter Boyce, then Executive Officer of the AAS, with funding from the NSF. An important feature of his program was that for most references a reader could click on the reference and the entire original paper appeared. That was made possible because the ADS (Astronomical Data Systems) has optically scanned all the major astronomical journals so that bitmaps could be brought up quickly and at no charge. That scanning was supervised by Guenther Eichhorn with large and continuing funding from NASA.

The on-line editions of astronomical journals quickly became very popular. By 2000 the ApJ sites (there are mirror sites in Europe and Japan) were receiving about 1 million visits per month (Abt 2001). Some astronomers state that they rarely go to their libraries because they can read the on-line editions of most astronomy journals in their offices. Most journals "bundle" subscriptions to the on-line editions with the paper editions and allow all astronomers and students to use the on-line editions free if their parent organization subscribes to the journals.

That raises the question of how journals are financed. Commercial journals receive all their funding from subscriptions. That makes some journals so expensive (up to $1 US per page) that many libraries cannot afford to subscribe. Some journals receive some organizational funding. The method used by most American non-profit journals is to charge the authors' institutions page charges that provide half to two-thirds of the journal income. Current rates are slightly above $100 US per page. That practice started for the ApJ as early as 1910 (Abt 1995b). The philosophy for page charges is that publication benefits both the authors and the readers, so both should share in the costs. Astronomy is a small field with small numbers of readers, so that advertising is not a practical source of income, as it is, for instance, in the medical field. The A&A practice is to charge Americans page charges but not astronomers in the European countries that support the journal through their national contributions. Many journals waive the page charges for astronomers in developing countries or those with serious financial problems.

8. Manuscript reviewing

It was mentioned when discussing observatory and conference publications that they had the disadvantage of no reviewing or unknown depth of reviewing. However, since fairly early in the century it has been the custom of nearly all astronomical journals to have reviewing by anonymous referees.

Astronomy and its literature have become so complex that we often overlook a pertinent paper or fact, and it is the role of the referee to point out omissions. He or she is also a critical reader who should point out weak or unclear arguments. With one-third of the authors and readers working in a language in which they are not native, the problem of making texts as clear as possible is always with us. The referee should help the author(s) to improve his or her manuscript. It rarely happens that a referee tries to prevent publication of ideas that he/she does not like or to delay publication while he/she completes similar work. Many journals have the policy that if the authors and referee do not come to an agreement after two rounds of reports and revisions, the manuscript is referred to another person for arbitration. Therefore a hostile referee cannot delay publication for a long time. Referees do their work, which is often very laborious, in exchange for having others help improve their papers.

There has been discussion recently as to whether it is fair for the referee to be anonymous. Or perhaps the authors should be anonymous to the referee. The latter is not practical because in the small field of astronomy where the equipment used is mentioned and the references often reveal the author(s), the referee can usually guess the identity of at least one of the authors. Regarding the reverse, it is feared that some authors may resent the recommendations of the referee and retaliate; again in this small field, requests for recommendations regarding grant requests or promotions might well be referred to the person who was upset by a negative referee report. Many senior astronomers who are less vulnerable to outside review of their positions or work will volunteer to be identified as the referee. We often recommend that younger astronomers who are building a career remain anonymous.

9. Electronic Tools

Indexes to journals have become increasingly important as the journals have grown to thousands of paper. Volume or annual indexes to authors and to subjects have appeared throughout the century in most journals. General Indexes every five years or so saves searching through individual volumes if the date of a remembered paper has been forgotten[1]. At first the indexes were compiled laboriously as typewritten manuscripts, but later the Press staff maintained a database in which the block of information of the title, authors, and locations for each paper was stored under each author's name and each subject heading. Now the General Indexes are available on-line.

The difficulty of having a separate index for each journal is that papers on a given topic or by a specific author must be searched for in each journal. However we now have web-based search engines that can find such papers in all journals. One problem with search engines is that most search for any specified words in texts, even if it involves only trivial comments. That results in the recovery of many unwanted sources. A search that was based only on key words would be more useful.

A very popular recent addition to the web is the preprint servers, such as the Los Alamos Preprint Server for astronomy and physics (Astro- Ph). Any author can submit his paper at any stage (between submission and publication) for listing and readers can call up new papers as they are added. The listings are not critiqued or removed from the listing, even after publication or rejection. Prudent authors wait to respond to referee reports before listing their papers. However this service provides a summary of current thinking, even if not all papers can be believed.

In many of the current journals the papers are transported via the web from authors to publishers to editors to referees, etc., and eventually to the presses for publication. They never have to wait for postal deliveries. Once they are accepted for publication, manuscript editing and composition takes only a week or so. The papers are then posted on the web in final form, except for the final pagination, in as little as 2-3 weeks after scientific acceptance. This is an important improvement over the previous 6-month publication time after acceptance.

An interesting benefit of the new current publication process is the ability to count citations, or references, to published papers. Beginning in 1955 the Institute for Scientific Information compiled a Science Citation Index in which is listed for every paper published in thousands of journals the other papers that make reference to it. Thus one can find for a given paper the names of the authors and locations of the papers that cite that paper.

This huge compilation, which occupies meters of shelf space each year, can be used in at least two ways. First is the scientific interest in knowing

[1]As a sidelight, this author found General Indexes so useful that when the ApJ stopped publishing them after Volume 100 in 1944, he volunteered to compile one for Volumes 101-135, then 136-145, and 146-165. When it came time to find a new Managing Editor to replace S. Chandrasekhar in 1971, this author's service to the ApJ was remembered and that is partly why he was selected for that position.

who is working on the topic of one's paper so that the author can read the later relevant papers. Second, the citation counts become evidence of the impact that the paper had on later work. A typical paper in astronomy receives 1-5 citations per year. The frequency of those citations usually increases to a maximum about five years after publication and then an exponential decrease ensues. A typical lifetime to half the peak citation counts is about 10 years. However important papers can be remembered for decades.

10. The Future

What can we predict for the near future? We have the means to distribute papers very rapidly after acceptance. But the technical production is still a laborious process that, if done with care, currently costs roughly $100 US per page. The organizations that produce those journals must recover their costs or they will cease publication. Should those costs come from the readers only or partly from the authors, institutions, or national governments? Once this financial problem is solved, distribution of papers can occur more widely.

The ADS has scanned all of the standard astronomical journals and is now scanning observatory publication and conference proceedings. The largest problem in scanning is that the bound volumes must have their spines cut so that individual pages can be fed through optical scanners. That usually means finding sets that will be unsuitable for further use. So in a few years nearly all the past worldwide astronomical literature will be available (free to all) on the web (normally until one year before the current issues).

Conference proceedings give the state of specialized fields and are primarily of current value. However their publication process takes an average of about 1-2 years. The delay is often because publication depends upon the slowest of the many authors to submit their manuscripts. If those proceedings were distributed on the web, rather than in book form, each paper could be posted as it is submitted. Further, because the time and funds of the editors are generally volunteered, there is neglible cost in such a form of publication. Librarians do not object to conference proceedings on the web; they will catalog them if they are in that form only. Thus the proceeding could be made available in a timely and inexpensive form. It is unlikely that a publisher will want to sell copies of books that are available free on the web, so both forms of publication probably will not occur.

The Europeans have learned that by combining their individual national astronomical publications into a combined "European Journal", it became very successful and now competes with the best journals. In the same way

I can visualize a single astronomical journal for eastern Asia, one for South and Central America, and one for the near-East.

On the other hand, if the use of search engines to find papers of interest becomes much more widespread and all journals are available on- line, it will not matter where papers are published; the readers will find and read them.

We mentioned that a problem with individual journal indexes is that they apply to only those individual journals. It should be possible to combine the indexes of all the astronomical journals into a single on-line index. Its advantage over search engines that scan everything is that it would be limited to refereed journals that have more reliable material.

There are plans for a "Virtual Observatory" that contains observational data of all kinds from many different locations. In the same way our published literature will become available to all astronomers, even to those whose institutions do not have large collections of their own. Astronomy in the future will become much more efficient and effective.

References

1. Abt, H.A. 1981, *Publ. Astron. Soc. Pacific* **93**, 269
2. Abt, H.A. 1990, *Publ. Astron. Soc. Pacific* **102**, 368
3. Abt, H.A. 1992, *Scientometrics* **24**, 441, Table 3
4. Abt, H.A. 1995a, *Publ. Astron. Soc. Pacific* **107**, 401
5. Abt, H.A. 1995b, *Astrophys. J.* **455**, 407
6. Abt, H.A. 2000, *Publ. Astron. Soc. Pacific* **112**, 1417
7. Abt, H.A. 2001, in *Astronomy for Developing Countries*, Ed. A.H. Batten (Provo, Utah: ASO Conf. Series), 354

THE EVOLUTION AND ROLE
OF THE ASTRONOMICAL LIBRARY AND LIBRARIAN

B.G. CORBIN
US Naval Observatory
3450 Massachusetts Avenue NW
Washington, DC 20392-5420, USA
corbin.brenda@usno.navy.mil

Abstract. Astronomy libraries and librarians have evolved as different techniques in access to information have become available. Both before and after the advent of the computer, librarians have been very active in getting information to the researchers as well as preserving the information for future generations, the main goals of the library and librarian.

1. Formation of the Astronomical Library

Most astronomical libraries came into being when an observatory or astronomical institution was founded, or a few years following the beginning of the institution. These libraries were established to accumulate reference materials for the staff and for the storage of information.

Observatories were founded mostly by either governments or universities and other educational institutions. Examples are the Paris Observatory, which was founded in 1667 with the library originating following a royal order of Louis XVI dated February 26, 1785, the Royal Greenwich Observatory Library, the Royal Observatory Edinburgh, and many other European astronomical libraries that were also founded in the 18th century.

Libraries in the United States and Canada were established at later dates, many in the 19th and 20th centuries. The Harvard College Observatory Library dates from 1839, while the US Naval Observatory (USNO) Library was established in 1844 and the Cincinnati Observatory Library was founded in 1845.

Owen Gingerich (1997) has described what he considered the five greatest astronomical collections of rare materials. These are the Royal As-

A. Heck (ed.), *Information Handling in Astronomy – Historical Vistas*, 139-155.

tronomical Society, London, the Paris Observatory, Pulkovo Observatory, St. Petersburg, the Uppsala University Observatory and the Royal Observatory, Edinburgh. Although he was focusing on the collections of books printed before 1800, the standard definition of a rare book, these libraries also form some of most complete collections of astronomical literature.

2. Astronomical Literature in Early Journals

Astronomical literature was of course published long before astronomical journals were established. In earlier times it existed as manuscript correspondence among astronomers, and later became a more generalized form of communication when these letters were sent to editors and publishers of early journals. Many astronomical papers appeared in the early journal *Acta Eruditorum*, and to a larger extent in the *Philosophical Transactions of the Royal Society, London.* Johann Hevelius reports observations made in Gedani, Poland, of the total eclipse of the moon which occurred on Feb. 21 and 22 in 1682 in Tom. 1 of *Acta Eruditorum*, 1682 (p. 109-116).

In the first issue of the first volume of the *Philosophical Transactions* dated March 6, 1665, the first three reports in this issue are on astronomical topics; "An Accompt of the Improvement of Optick Glasses at Rome", "Of the Observation made in England, of a spot in one of the Belts of the Planet Jupiter", and "Of the motion of the late Comet predicted". These accounts were based on letters written to the editor of the *Phil. Trans.* to report results of astronomical research. Many astronomical papers covering all aspects of astronomical endeavor also appeared in the publications of the scientific societies of various countries.

A long series of papers by French astronomers appeared through the years in the *Mémoires de l'Académie Royale des Sciences de l'Institut de France.* Most early astronomical libraries received these journals, as well as the scientific society publications from most European countries. These journals and series formed the periodicals sections of the libraries.

3. Published Library Catalogs

Making the astronomical literature available to the staff of an observatory was, and is, the primary goal of the astronomical library. In order to also share this information with astronomers outside an institution, some libraries published their catalogs in book format. Some appeared in observatory publications such as those of the Lick Observatory in 1891 and the Woodman Astronomical Library of the Washburn Observatory, University of Wisconsin in 1884.

There were several separate catalogs published in book form: in 1886 for the Royal Astronomical Society, in 1890 for the Crawford Library of the

Royal Observatory Edinburgh, and in 1878 for the Library of the Observatoire Royal de Belgique, Brussels. The *Catalogue of the Crawford Library* (2001) has very recently been reprinted in a limited facsimile edition. The *Catalog of the US Naval Observatory Library* (1976) resulted from the microfilming of the catalog cards by the publisher, G.K. Hall.

As the USNO library has an extensive collection, the publisher felt the catalog would serve as a bibliographic reference for astronomical literature. These earlier printed catalogs were the forerunners of the online catalogs developed in the 20th century. The focus of these early catalogs was the same as at present, to let users outside an institution know the publications held in that library.

4. Card Catalogs

The major way that astronomical literature in a collection was available to the researchers was via the card catalog. In the age before computers, the card catalog was a very labor-intensive part of the library's task. Cards were prepared, handwritten in the early days, typewritten later, with a separate card being prepared for each component of the description of the book or journal; author, co-author(s), title, series, and subject terms. One book could require as many as ten cards with the basic information the same but the top line for filing the card being different.

Many libraries spent large amounts of time preparing analytics, or cards listing titles of important journal articles of particular interest to an observatory. Thus, the analytics would differ from library to library. Catalog cards contained a wealth of information. By browsing the card catalog, researchers would often encounter important works they might not have been familiar with otherwise. Each library usually prepared a shelf list which was the librarian's working copy of cards filed in call number order. Important notes were often placed on these cards such as the astronomer who requested the book, the price paid, whether the book was missing from the shelf, and even sometimes a note as to where the book had been reviewed.

Nicolas Baker (1994), in a controversial article which appeared in the *New Yorker Magazine*, discussed in depth what he views as a tragedy when complete card catalogs were disposed of after they were put into computer form for the online catalog. Not all of the information on the cards was put into the online catalog, and therefore in many cases important details were lost. He felt in such instances complete histories of the libraries' collection was now lacking.

The US Naval Observatory Library has determined, at least for the present, to retain its card catalog and shelf list although new cards are no longer being filed in the main catalog. In the 19th and early 20th centuries, many astronomy libraries would also keep manuscript books in which they entered "accessioned items" often assigning a running accession number to the items, and listing the dates the books were received, from whom they were purchased or received on exchange, price paid, etc. It is fascinating to look at the older accession books to get a feel for the development of a library's collection.

5. Classification Schemes Used in Astronomy Libraries

Astronomy libraries use various systems for classifying materials and these have changed through the years. Many observatories developed their own schemes which were then used by other libraries.

Edward S. Holden developed a classification system that was used at the US Naval Observatory Library for many years. However, as standard systems became more prevalent, many astronomy libraries adopted these systems. At present, many libraries in the US use the Library of Congress classification system while some US astronomy libraries use the Dewey Decimal Classification System.

A number of other classification schemes are used in the rest of the world. Many libraries in Europe and India use the Universal Decimal Classification (UDC) and those in South America use a variety of schemes including Dewey, UDC and in-house classification systems.

One interesting example of an in-house system in use is the Dewhirst Classification at the Institute of Astronomy in Cambridge, UK. David Dewhirst, an astronomer who was also the librarian developed this classification based on the identifying numbers used in *Astronomy and Astrophysics Abstracts.*

6. Archives

The observatory archives are often placed under the care of the library in many institutions and librarians have produced finding aids for use with the archives. One example of an extensive, well-cataloged collection is the Mary Lea Shane Archives of the Lick Observatory. However, in other libraries there is not sufficient staff to catalog and care for both the library collection and the archival collection, and sometimes the archival collection receives less attention due to this lack of staff. In some instances, the observatory's plate archives also have become the responsibility of the library.

At Lowell Observatory, the library has undertaken an extensive project to place all plates in acid free envelopes, transfer the information from the older envelopes to the new envelopes and also place this information online to form an online catalog of the plates.

Figure 1. The reading room of the Pulkovo Observatory library.

Figure 2. The US Naval Observatory library.

Of interest is the recent establishment of the facility being developed at the Royal Observatory of Belgium in Uccle-Brussels (De Cuyper *et al.* 2001). This center hopes to collect plate collections from observatories around the world in order to preserve, catalog and later digitize these plates. If sufficient funding is available for this project, it could prove to be the solution to plate storage and preservation problems at many observatories.

Some libraries have the historical photograph collections within the library. Yerkes Observatory, the US Naval Observatory, and Lowell Observatory have begun projects to scan and place these photographs online. Similar projects are taking place in Italy and Finland, and projects for scanning photographs and historical archives are being planned in India.

7. Librarians

In the earliest astronomical libraries, an astronomer would often be placed in charge of the library. This tradition continued for many years until the early to mid-1900s when a transition to professional librarians began. However, the tradition of the astronomer being in charge still continues in some university astronomy department libraries, where the collection is small and funds are not available for a librarian position.

It is not unusual for an astronomy librarian to now spend his or her complete career in the same library. This is very beneficial for the library for many reasons. The person becomes very familiar with the complete subject area of astronomy and its literature and also develops into a subject specialist on the type of research being done at the parent institution.

This knowledge allows the librarian to be on the lookout for specific books and articles which would be immediately useful for staff members. Astronomy librarians have the opportunity to concentrate on one subject specialty while their colleagues in science libraries at universities sometimes must cover three or four subject fields.

8. Communications among Librarians

The world of astronomy librarians has always been small and interconnected, even before the advent of the computer and email. Librarians communicated via postal mail for varying reasons such as seeking missing observatory publications or missing journal issues. At times librarians or astronomers would find publications listed in the *Astronomische Jahresbericht* which had not been received in the library but were needed by researchers. Communications over the years would often be with the same librarians, so a feeling of kinship arose, even though the librarians had never met.

Another example is the attempts of libraries to obtain issues of journals or observatory publications issued during World War I and II which had either never been sent, or had been lost in transit. The US Naval Observatory Library only completed filling in all missing issues of the *Astronomiche Nachrichten* in the 1970s, and some of these issues were photocopies very graciously copied at Hamburg as original issues were no longer available. This filling in of missing issues even extended to the Cold War era when it was sometimes difficult to exchange materials. Thus, the close connections among librarians have always been a benefit to the staff in maintaining a collection which was as complete as possible.

9. Astronomers and Librarians

Astronomers have a dependence on the librarian to make needed information accessible. Before the advent of the computer, for example, the librarian would make absolutely sure that all indexes had been received and were present when a volume was bound. Before computers there was a total dependence on the index to a volume as it was essential to find the authors or subjects of papers in that volume. The librarian was required to support the astronomical research by keeping current with books being published in the field.

Not as many books were published in earlier times, but it was not always easy to find addresses of foreign publishers or find the current prices. Much time was needed to cover ordering problems. The fact that astronomy librarians stayed in their positions for long periods made some of these tasks a bit easier. Librarians have continued to respond to the needs of astronomers both through new acquisitions and new electronic resources. In the last twenty years this has included being aware of new databases, interfaces, and new ways of exchanging and pointing to information.

As André Heck (1993) has pointed out, "There is now a new generation of librarians very active in our community, 'new' being not a question of age, but representative of these new attitudes towards information retrieval and of various remarkable initiatives and undertakings." Most recently librarians have begun to collaborate with astronomers from other institutions on a wide scale under the auspices of Commission 5 of the IAU. This is discussed in more detail later.

10. Observatory Publications

Most observatories initially distributed results of research at their institutions in the form of observatory publications. These began in the 18th century and continued until the early 1990s in some cases. Some observatories still issue publications, but it is much more infrequent at present. These series of publications from observatories worldwide contain information on every facet of astronomy. For example, these are often the only places where one finds detailed descriptions of astronomical instruments of different eras. The publications were sent on an exchange basis to most observatories.

The US Naval Observatory's collection, which is almost complete, contains approximately 3000 volumes in these series. Articles in the publications differ from journal articles in several ways. The observatory publication articles were often much longer than journal articles and reported in detail on the observatory's projects. Often star catalogs, including variable star, double star, and astrometric catalogs, would appear here. Cometary and galaxy catalogs, parallax data, and solar studies were other types of information appearing. These articles reported both on intermediate results and finished projects. These series constitute important collections of data in astronomy which are preserved in libraries.

The author has long pushed for a project which would preserve these publications. This is now taking place, and will be described later in this chapter.

11. Libraries as Historical Databases

Libraries are preservers of historical data. Many collections have large numbers of older volumes on the shelves which many people initially thought were no longer useful. However, the role of the library is to preserve the historical record of the subject in addition to providing current information. There is now clear evidence of the importance of this preservation role in recent projects which have used these older data. Minor planet work and planetary ephemerides computations require early observations for compilations that produce modern orbits or catalogs.

As early star catalogs were put into electronic format, these catalogs were sent to CDS in Strasbourg and to other data centers throughout the world. The data from these catalogs made possible such modern catalogs as Tycho II, IRS, ACRS and the PPM. Various observatories depended on their libraries to have these earlier publications available to make these projects possible. The Astrographic Catalog (AC) is another example. Modern reductions of the AC are now used by astronomers worldwide, a century after many of the volumes appeared. Double star projects also show the use of older data. From the historical volumes held and cataloged by librarians, with many of the double star observations appearing only in journals, thousands of double star observations have been compiled into large databases. At present many orbits have been determined due to the availability of these data in libraries.

Astronomers rely on librarians to interpret obscure and sometimes incorrect references to double star, variable star and other catalogs in searching for the older data. Other areas of astronomical research that have benefited from the accessibility of older data are solar research, variable stars, eclipse information and cometary studies.

The *Astronomische Jahresbericht* and later *Astronomy and Astrophysics Abstracts* listed all publications in astronomy and researchers could find the publications of interest to them in these volumes. It is important to note that not only were these materials preserved by the libraries, but also they were cataloged and indexed so that all the references could be found for these modern projects.

12. History of Astronomy

Historians of astronomy and researchers looking at the history of various astronomical topics rely on library collections. Interaction occurs among the historians and librarians as these projects unfold. As historians discuss their findings, librarians have a clearer picture of where to find out of the way, and perhaps hidden, historical information in works of the period. Again, observatory publications contain much historical information about instruments, buildings, long-term observing programs and other astronomical projects.

Libraries preserve manuscripts and photos which are invaluable to the historian. Materials for current histories of modern projects are often found in the library as the librarian catalogs even small brochures, in some cases press releases, describing an observatory's new instruments. However, historical astronomy is one area where obviously everything is not on the web. A historian recently stated that it is a lot easier than even a short five years ago to find historical astronomy material on the web, but it will take some time before all the significant bibliographical information on the history of astronomy finds its way to the Internet.

Future historians will still want and need to have some knowledge of what is recorded in the older written records. Another loss for the history of astronomy is those library collections which have been damaged by fire, flood or other loss. One example is the collection of the Pulkovo Observatory Library which suffered severe loss from a fire set by arsonists in 1997.

13. Merging of Collections

Many departmental astronomy libraries have been absorbed into larger university science libraries. In some cases, due to space concerns, the collection has been "weeded" or parts of the collection discarded. This is especially true of volumes which might not have high use such as older volumes and observatory series. This makes it very important for larger observatory libraries to keep everything relating to astronomy for historical reasons.

Some librarians, usually with the assistance of some staff astronomers, have been required to strongly defend their collections to over-zealous heads of institutions who wish to discard parts of the collection, usually older or presumed out of date books. Often a need for more office space is the reason given for these "toss out" decisions.

In many cases where the departmental astronomy library has been integrated into the larger science library, astronomers find it more difficult to easily consult the literature they need. Some core journals, or at least recent years of these journals, stay in the astronomy department. One must sometimes visit two other campus libraries to access everything that was once in the astronomy library. This problem of access has been around quite a while.

In an unpublished survey of observatory libraries in the United States taken by Mary T. Howe (1940), one question asked where materials were kept, in the main library or the astronomy department. One astronomer replied, "all the books and purchased periodicals are in the main library of the university where they are conveniently accessible to all except those who need to use them". Another problem facing astronomy libraries is the necessity to store older materials off campus, requiring a day or more to retrieve the material needed.

The false statement that "everything is on the web" continues to be repeated, and the astronomical researcher continues to need access to older materials not yet available electronically.

14. Astronomy Librarian Groups

The International Federation of Library Associations (IFLA) formerly had a subsection called The Astronomical Society and Observatories Libraries, which was formed in 1965. European librarians were active in this subsection. G. Feuillebois, Librarian of the Paris Observatory, was the first president of this group which held its first meeting at The Hague in 1966. This group, under her leadership, was active in planning projects.

An important one was the publication of the list of *Non-commercial Publications of Observatories and Astronomical Societies* published in four editions from 1971-1981 by the Sonnenborgh Observatory Library at Utrecht. The *Union List of Astronomical Serials*, discussed later in the chapter, was a follow-on of this work. When IFLA held its international meeting in Washington, DC in the early 1970s, this subgroup of astronomy librarians held sessions at the conference. A few US librarians joined in this meeting, and a special tour of the US Naval Observatory Library was planned. The author met these astronomy librarians face to face for the first time. The group became less active in later years, and to my knowledge this subsection of IFLA no longer exists.

The Physics-Astronomy-Mathematics Division (PAM[1]) of the Special Libraries Association (SLA), was formed in 1972. Although this group consists mainly of US and Canadian librarians, in recent years European, Asian, and South American librarians have joined PAM. The PAM Division has been active, and the subset of astronomy librarians has taken on an active international role. One very useful project sponsored by PAM was the compilation of the *Union List of Astronomy Serials* (ULAS[2]). Judith Lola,

[1] `http://pantheon.yale.edu/~dstern/pamtop.html`
[2] `http://sesame.stsci.edu/lib/union.html`

librarian at the Yerkes Observatory was the compiler and SLA published the first edition in 1983. The second edition, available only online, was updated by Judith Lola Bausch and Sarah Stevens-Rayburn, librarian at the Space Telescope Science Institute, with holdings of additional observatory libraries added.

The PAM group instituted an International Librarian Award in 1998 in which a librarian from a developing country is given a two-year membership to Special Libraries Association and the PAM Division. Although not specifically planned this way, astronomy librarians have been chosen for this award three of the four times it has been given.

15. IAU Commission 5 and Libraries

The Commission 5 of the International Astronomical Union, Documentation and Astronomical Data, has long cooperated with libraries. There is a Working Group (WG) on Astronomical Libraries, which has been active at times with inactivity in other periods. G. Feuillebois, who was a Co-Chair of this group in the 1960s, attended the XIIIth IAU General Assembly in Prague in 1967 and reported to Commission 5 on possible projects. After this period, the WG became inactive for many years.

The author attended the XVIIIth IAU General Assembly in Patras, Greece in 1982 and was later made a consulting member of the Commission. George Wilkins, then President of Commission 5, wished to have the libraries WG activated again and asked the author to become co-chair of the WG, a position held for three of the IAU triennial terms. Uta Grothkopf, librarian of the European Southern Observatory (ESO) is the current co-chair of the libraries WG.

Commission 5 has always been supportive of library projects and provides an important link between astronomical librarians and professional astronomers. In the early 1990s, Robyn Shobbrook, then librarian of the Anglo-Australian Observatory (AAO) and Robert Shobbrook, an astronomer at AAO, compiled the *International Astronomical Union Thesaurus*. This work, to which many librarians contributed, was carried out under the sponsorship of Commission 5 (Shobbrook & Shobbrook 1992).

At present, a new WG on Electronic Publishing has been established, and 2 librarians have been asked to join this group. Marlene Cummins, librarian of the Department of Astronomy at the University of Toronto, and Uta Grothkopf, ESO, are the current representatives.

16. LISA Conferences

The first Library and Information Services in Astronomy (LISA) conference was held in Washington, DC in 1988 and was the first full-scale meeting of astronomy librarians from all parts of the world. A brief history of how the LISA conference came into existence is perhaps of interest. When the author attended the XVIIIth IAU General Assembly in Patras in 1982, she asked if she might present to Commission 5 a brief overview of projects of interest to astronomy librarians. She proposed that an international meeting of astronomy librarians, planned and sponsored by the Commission, be taken under consideration.

The group decided to support the idea of a meeting, and the author continued to discuss such a meeting with other astronomy librarians and with Gart Westerhout, at the time the Scientific Director of the US Naval Observatory, and also a member of Commission 5. The author did not attend the XIXth IAU General Assembly in Delhi in 1985, but asked Gart Westerhout to continue to discuss the possible conference with the group. He returned from Delhi and reported that Commission 5 had warmly welcomed such a meeting and would ask for financial support from the IAU. He also reported that he volunteered the US Naval Observatory to act as host for the conference and volunteered the librarian, the author, to do the planning!

The first LISA conference was held in Washington, DC in the summer of 1988, just before the XXth IAU General Assembly in Baltimore, MD. It was a very successful conference with approximately 125 librarians and astronomers in attendance from all parts of the world. It was a very exciting time as librarians met face to face with colleagues they had corresponded with for many years. The close contacts among astronomy librarians always existed, but the contacts became more formalized with the LISA meeting. The internet announcement list, ASTROLIB, was an outgrowth of LISA I. Ellen Bouton, librarian at the National Radio Astronomy Observatory USA, manages the list. Librarians send messages to Ellen Bouton, which are then sent to astronomy librarians worldwide. It has proven to be a very successful means of communication among all astronomy librarians.

LISA II was held in Garching, Germany in 1995, sponsored by the European Southern Observatory, LISA III in Tenerife, Canary Islands in 1998, sponsored by the Instituto de Astrofísica de Canarias, and LISA IV in Prague in 2002, sponsored by the Astronomical Institute of the Charles University, Prague, and the Astronomical Institute of the Academy of Sciences of the Czech Republic. The proceedings of the first three LISA conferences have been published:

LISA I – Wilkins & Stevens-Rayburn (1989),
LISA II – Murtagh *et al.* (1995) and
LISA III – Grothkopf *et al.* (1998).
The proceedings of LISA IV are in press, Corbin *et al.* (2002).

LISA brings librarians into contact with scientists from other institu-

Figure 3. J. Watson (center) with W. Warren (right) and A. Heck at the 169th American Astronomical Society Meeting in Pasadena (January 1987) standing in front of the historically first demontration booth for the CDS database Simbad at such conferences. Joyce became the "face" of SIMBAD in the US for quite a number of years.

tions. Before, librarians had contacts with scientists at their own observatories, but not so many from a broad range of institutions. LISA provides cross-pollination and more contact. One result is that librarians now collaborate with other librarians and astronomers in international projects. As a result of this collaboration, two librarians have become full members of the IAU.

As the LISA meetings have been very successful, one would hope they would continue to take place perhaps every four years as they provide an opportunity for astronomy librarians and astronomers to share their ideas and discuss future projects.

17. Introduction of Computers into Libraries

The Library of the Royal Observatory Edinburgh (ROE) was perhaps one of the first to use machine-readable techniques in a library application. D.A. Kemp (1967), ROE Librarian, published a report describing how they produced a catalogue of their current working collection on paper tape via a Flexowriter. This allowed them to print the catalogue either on index cards or in page format, plus retain the titles in machine-readable form.

In the 1980s, many US libraries began to catalog their collections on OCLC[3], a national bibliographic database. The data could be downloaded onto magnetic tapes and become the basis for an online catalog. Also, in the late 1980s, some libraries first began to have access to email; for example, email came to the US Naval Observatory in 1987.

In the conference "Weaving the Astronomical Web" held in Strasbourg in April 1995, Jane Holmquist (1995), librarian of the Astrophysics Library at Princeton University reported on the continuing development of their library web page which gave the staff access to astronomical electronic resources. Laura Abrami and A. Balestra (1995) reported on the computerization of the Astronomical Observatory of Trieste Library.

Joyce Watson (1922-2001), then librarian at the Smithsonian Astrophysical Observatory, Cambridge, Massachusetts, was perhaps the librarian best known for leading the way into the new world of online databases. Her article, "Astronomical Bibliography from Commercial Databases" (Watson 1991), gave a thorough introduction to the new world astronomy librarians were facing.

18. Current Status of Computer Catalogs and Other Projects

Most astronomy libraries now have their catalogs online, and many are available on the web so users around the globe can search these catalogs. This has been very beneficial to researchers as it provides access to libraries which have very complete collections in astronomy. This puts information not easily available in earlier years into their hands very quickly, usually via an email request to the library, and a fax that can be sent immediately if the request does not involve a large number of pages.

Libraries such as the Paris Observatory and Paris-Meudon Observatory have been very active in cataloging even the very small pamphlets and other not widely known astronomy items on the OCLC database. The US Naval Observatory has received several grants from the Legacy Program of the US Department of Defense to do online cataloging on OCLC of the rare book collection, parts of the 19th century collection, and individual

[3] `http://www.oclc.org/home/`

articles appearing in observatory publication series. Researchers using the very widespread WorldCat database on OCLC can easily find these publications by searching either author or title, and WorldCat then gives the holding libraries of the publications.

The role of the library in making information available thus has been widened and expanded, both by giving its staff access to other collections as well as making information available to the world at large. Some other projects planned and carried out by librarians include online lists of astronomy newsletters, annual reports of observatory publications online, calendars of astronomical meetings, and a list of book reviews of current astronomy books.

19. The Astrophysics Data System and Libraries

In the early 1990s, NASA formed a pilot project called STELAR (STudy of Electronic Literature for Astronomical Research). Astronomers, librarians and computer specialists joined to explore the possibility of having full text documents on line for searching. It was the effort in this pilot project on which the Astrophysics Data System[4] (ADS) was founded.

It is probably safe to say that the ADS, has revolutionized the way astronomers do their research. Joyce Watson contributed ideas to the ADS from the very beginning and made suggestions about indexing from a librarian's point of view. These suggestions were incorporated early on making the ADS an excellent product right from the beginning. Librarians suggested to the ADS that they strongly consider not just including the current literature, but scan astronomical journals back to the beginning volume if at all possible.

Great credit should be given to Guenther Eichhorn[5] of ADS for carefully listening to this suggestion and agreeing to make this inclusion. Librarians have assisted by donating long runs of duplicate copies of older journals to be scanned and be placed full text online. It has been quite surprising and amazing to see how heavily the older literature has been used. As the ADS has a number of mirror sites around the world, most astronomers can easily access this important resource.

Since 1977, the author has pressed for the preservation of the long runs of observatory publications. Success seemed at hand on several occasions (Corbin & Coletti 1995), but things never quite worked out. However, Donna Coletti, librarian at the Harvard-Smithsonian Center for Astrophys-

[4]`http://adswww.harvard.edu/`

[5]See the chapter by G. Eichhorn *et al.* on *The Development of the Astronomy Digital Library* in this volume. (Ed.)

ics, listened to the reasons why this preservation was necessary and became an active proponent for the project.

With grants from the US National Endowment for the Humanities, Harvard University undertook a large-scale preservation-microfilming project of scientific collections, including astronomy. Many complete series of observatory publications have been microfilmed, and the project is still underway. Copies of the microfilm are now being digitized and the full text placed on ADS.

Therefore, this preservation project has become a much more encompassing one, making the full text of important data published by observatories from all over the world from the late 1700s to the early1990s available to researchers around the globe. Guenther Eichhorn and Donna Coletti are owed a great debt of thanks by the astronomical community.

20. The Future

It is clear that astronomy libraries are not just places to keep up with the current literature. As more of the literature becomes available online, researchers will have it at their fingertips. It is clear that perhaps not all literature will be available, but librarians should anticipate the explosive growth in information storage and electronic retrieval capacities and press for preserving most of the astronomical literature, including photographs.

The ADS has already made a wonderful start towards this end, and with proper funding could expand even further the amount of astronomical literature available. The task of an astronomy library will continue to be making the literature available first of all to their staffs, but ultimately to a worldwide staff as well.

Librarians will continue participating in the development of information systems and cooperating with astronomers in making sure the data are always at hand.

References

1. Abrami, L. & Balestra, A. 1995, Offering a Library on the Internet: the OATs Experience, in *Weaving the Astronomical Web*, D. Egret & A. Heck (eds.), *Vistas in Astron.* **39 (1)**, 53-61.
2. Baker, N. 1994, Discards, *New Yorker Magazine*, **LXX (7)**, 64-70, 72-76, 78-86.
3. Catalog of the Naval Observatory Library, Washington DC 1976, G.K. Hall, Boston, 6v. (ISBN 0816100314).
4. Catalogue of the Crawford Library of the Royal Observatory, Edinburgh, Facsim. ed. Astronomy Books, Bernardston MA, viii + 499 pp. (ISBN 1578980127).
5. Corbin, B. & Coletti, D. 1995, Digitization of Historical Astronomical Literature, in *Library and Information Services in Astronomy II*, F. Murtagh, U. Grothkopf & M. Albrecht (eds.), *Vistas in Astron.*, **39 (2)**, 161-165.
6. Corbin, B., Bryson, E. & Wolf, M. (eds.) 2002, *Library and Information Services in Astronomy IV*, US Naval Observatory, Washington DC, in press

7. De Cuyper, J.P. *et al.* 2001, The Uccle Direct Astronomical Plate Archive Centre UDAPAC-A New International Facility for Inherited Observations, in *Astronomical Data Analysis Software and Systems X*, F.R. Harnden, Jr., F.A. Primini & H.E. Payne (eds.), *Astron. Soc. Pacific Conf. Series* **238**, 125-128.
8. Gingerich, O. 1997, The World's Greatest Rare Astronomy Libraries, AB Bookman's Weekly, Vol. **100** no. 17 (October 27), 1022-1028.
9. Grothkopf, U., Andernach, H., Stevens-Rayburn, S. & Gomez, M. (eds.) 1998, Library and Information Services in Astronomy III, *Astron. Soc. Pacific Conf. Series* **153**, xxxii + 323 pp. (ISBN 1886733732).
10. Heck, A. 1993, The increasing role of librarians in astronomical information retrieval, *Bull. Inf. Centre Données Astron. Strasbourg* **42**, 51-55.
11. Holmquist, J. 1995, Library Services and the Web, in *Weaving the Astronomical Web*, D. Egret & A. Heck (eds.), *Vistas in Astron.* **39 (1)**, 47-51.
12. Howe, M.T. 1940, Astronomical Observatory Libraries. Preliminary edition, May 1940, New York City. From an unpublished mimeographed typescript in the US Naval Observatory Library
13. Kemp, D.A. 1967, Astronomical Literature – Available Services and Possible Developments (based on a talk given to the Royal Astronomical Society 1967 Sep 8), Royal Observatory, Edinburgh, November 1967.
14. Murtagh, F., Grothkopf, U. & Albrecht, M. (eds.) 1995, Library and Information Services in Astronomy II, *Vistas in Astron.*, **39 (2)**, vi + 127-286 pp.
15. Shobbrook, R.M. & Shobbrook, R.R. 1992, The International Astronomical Union Thesaurus, Anglo-Australian Observatory for the IAU, Epping NSW, viii + 115 pp.
16. Watson, J.M. 1991, Astronomical Bibliography from Commercial Databases, in *Databases and On-line Data in Astronomy*, M. Albrecht & D. Egret (eds.), Kluwer Acad. Publ., Dordrecht, 199-210.
17. Wilkins, G.A. & Stevens-Rayburn, S. (eds.) 1989, Library and Information Services in Astronomy, International Astronomical Union Colloquium 110, US Naval Observatory, Washington DC, xv +244 pp.

THE DEVELOPMENT
OF THE ASTRONOMY DIGITAL LIBRARY

G. EICHHORN, A. ACCOMAZZI, C.S. GRANT,
M.J. KURTZ AND S.S. MURRAY
Harvard-Smithsonian Center for Astrophysics
60 Garden Street
Cambridge, MA 02138, USA
gei@head-cfa.harvard.edu

Abstract. The Astronomy Digital Library provides access to astronomical literature and to on-line data. The Astrophysics Data System (ADS) is the literature search system and archive in this library. It is a NASA-funded project and access to all the ADS services is free to everybody world-wide. The ADS Abstract Service allows the searching of four databases with abstracts in Astronomy, Instrumentation, Physics/Geophysics, and the ArXiv Preprints with a total of over 2.9 million references. The system also provides access to reference and citation information, links to on-line data, electronic journal articles, and other on-line information. The ADS Article Service contains the full articles for most of the astronomical literature back to Volume 1. It contains the scanned pages of all the major journals (Astrophysical Journal, Astronomical Journal, Astronomy & Astrophysics, Monthly Notices of the Royal Astronomical Society, and Solar Physics), as well as most smaller journals back to Volume 1. There are now 10 mirror sites of the ADS available in different parts of the world to improve connectivity. The ADS can be accessed through any web browser without signup or login at: http://ads.harvard.edu/ .

1. Introduction

Over the last 10 years, the astronomical community has developed an interlinked Astronomy Digital Library. The NASA Astrophysics Data System Abstract Service is by now a central facility of this library. In a typical month it is used by ~60,000 individuals, who make ~1 million queries,

A. Heck (ed.), Information Handling in Astronomy – Historical Vistas, 157-182.

retrieve ~50,000,000 bibliographic entries, read ~800,000 abstracts and ~130,000 articles, consisting of ~1,100,000 pages. The ADS is a key element in the emerging digital information resource for astronomy, which has been dubbed Urania (Boyce 1996). It is tightly interconnected with the major journals of astronomy, as well as with the major data centers in a connected system that is unique in sciences. A detailed description of the ADS has been published in a special issue of Astronomy & Astrophysics Supplements in April, 2000 (Overview: Kurtz *et al.* 2000; Search Engine and User Interface: Eichhorn *et al.* 2000; System Architecture: Accomazzi *et al.* 2000; Data: Grant *et al.* 2000).

This article will describe the history of the Digital Library in Astronomy and in particular describe the Astrophysics Data System, its data holdings, and its role in the bibliographic research in Astronomy.

2. History

The Astronomy Digital Library started with the ADS implementing a connection between the SIMBAD database at the CDS and the bibliographic database in the ADS. This allowed combined queries to two different on-line systems. Since then the Digital Library has grown considerably and includes many interlinked resources. The central literature service is provided by the ADS.

The first major part of the Digital Library is the literature search system. The ADS Abstract Service allows the searching of most of the literature in Astronomy and significant part of the Physics literature. It was started in 1993 with a custom-built networking software system to provide access to distributed data (Murray *et al.* 1992). By summer 1993 a connection had been made between the ADS and SIMBAD (Set of Identifications, Measurements and Bibliographies for Astronomical Data – Egret *et al.* 1991) at the Centre des Données de Strasbourg (CDS), permitting users to combine natural language subject matter queries with astronomical object name queries (Grant *et al.* 1994). The user interface for this first version of the ADS was built with the custom-built software system that the ADS used at that time. The search engine of this first implementation used a commercial database system. A description of the system at that time is in Eichhorn (1994a).

By early 1994, the World Wide Web (WWW 1999) had matured and was widely accessible through the NCSA Mosaic Web Browser (Schatz & Hardin 1994). It now was possible to make the Abstract Service available via a web forms interface, which was released in February 1994. Within five weeks of the initial WWW release, use of the Abstract Service quadrupled (from 400 to 1600 users per month), and it has continued to rise ever since (Eichhorn 1997a). The WWW interface to the ADS is described by Eichhorn *et al.* (1995a&b).

The second major part of the Digital Library is the archive of the full text of the astronomical literature. It contains scanned full journal articles for most of the astronomical journal literature going back to Volume 1. The first full article bitmaps, which were of Astrophysical Journal Letters (ApJL) articles, were put on-line in December 1994 (Eichhorn *et al.* 1994b). By the summer of 1995 the bitmaps for the ApJL were current and complete going back ten years. At that time the Electronic ApJ Letters (EApJL) (Boyce 1995) went on-line. From the start the ADS indexed the EApJL, and pointed to the electronic version. Also from the beginning the reference sections of the EApJL pointed (via WWW hyperlinks) to the ADS abstracts for papers referenced in the on-line articles.

With time, other interfaces to the abstracts and scanned articles were developed to provide other information providers the means to integrate ADS data into their system (Eichhorn *et al.* 1996b).

With the adoption of the WWW user interface and the development of the custom-built search engine, the current version of the Abstract Service was basically in place. Currently the ADS system consists of four semi-autonomous (to the user) abstract services covering Astronomy/Planetary Sciences, Instrumentation, Physics, and Astronomy Preprints. Combined there are over 2.9 million abstracts and bibliographic references in the system. The Astronomy Service is by far the most advanced, and accounts for ~85% of all ADS use (Kurtz *et al.* 2000; Eichhorn *et al.* 2000).

3. Data

This section describes the data holdings in the abstract and article service of the ADS as well as the links database that is the basis for the Digital Library.

3.1. ABSTRACTS

The abstracts in the ADS come from many different sources (see Grant *et al.* 2000). The original set came from the NASA STI database. We now receive basic bibliographic information (title, author, page number) from essentially every journal of astronomy. Most publishers also send us abstracts, while some who cannot send abstracts, allow us to scan their journals. For these journals we build abstracts through optical character recognition (OCR). Finally we receive abstracts from the editors of conference proceedings, and from individual authors.

As of Sept, 2002 there are ~825,000 astronomy references indexed in the ADS, the database is nearly complete for the major journals articles.

In the Physics database there are ~1.3million references, and in the Instrumentation database there are ~640,000 references. Approximately half of all references have abstracts, the other half only have titles, authors, and journal information.

Usage statistics of the ADS show that 85% of the queries are for authors or words in the title. This means that table of contents-only entries (without abstracts) still satisfy a majority of the queries.

3.2. BITMAPS

The ADS has obtained permission to scan, and make freely available online, page images of the back issues of all the major journals and most smaller journals in astronomy. In most cases the bitmaps of current articles are put on-line after an embargo period, to protect the financial integrity of the journal.

We plan to provide for each collaborating journal, in perpetuity, a database of page images (bitmaps) from Volume 1 Page 1 to the first issue which the journal considers to be fully on-line as published. This will provide (along with the indexing and the more recent archives held by the journals) a complete electronic digital library of the major literature in astronomy.

On a longer term we plan to scan old observatory reports and defunct journals, to finally have a full historical collection on-line. This work is beginning with a collaboration with the Wolbach Library at the Harvard-Smithsonian Center for Astrophysics and the Harvard Preservation Project (Eichhorn *et al.* 1997b; Corbin & Coletti 1995).

By now there are ~2 million scanned pages on-line in the ADS in ~280,000 articles. The bitmaps in the ADS have been scanned at 600 dpi using a high speed scanner and generating a 1 bit/pixel monochrome image for each page (see Grant *et al.* 2000). The files created are then automatically processed in order to de-skew and center the text in each page, and add a copyright notice at the bottom of each page. Adding the copyright notice on each page is important, since the ADS makes it very easy to reprint individual pages. Such individual pages would lose the information on where they came from and who owns the copyright for them.

3.3. LINKS

The system of links from the ADS to other on-line resources is the center part of the Astronomy Digital Library. It allows the user to easily navigate between different on-line information providers.

The ADS responds to a query with a list of references and a set of hyperlinks showing what data are available for each reference (see Eichhorn *et al.* 2000). There are ~6.5 million hyperlinks in the ADS, of which ~40% are to sources external to the ADS project.

The largest number of external links are to SIMBAD, the NASA Extragalactic Database (NED), the Space Telescope Science Institute (STScI), and the electronic journals. A rapidly growing number, although still small in comparison to the others, are to data tables created by the journals and maintained by the CDS and the Astronomical Data Center (ADC) at Goddard. These links are an extremely important aspect of the ADS.

Table 1 shows the links that we currently provide when available.

A more detailed description of resources in the ADS that these links point to is provided in Grant *et al.* (2000).

Some of these links (for instance the 'D' links) can point to more than one external information provider. In such cases the link points to a page that lists the available choices of data sources. The user can then select the more convenient site for that resource, depending on the connectivity between the user site and the data site.

3.4. CITATIONS AND REFERENCES

The use of citation histories is a well-known and effective tool for academic research (Garfield 1979). In 1996 the American Astronomical Society purchased a subset of the Science Citation Index from the Institute for Scientific Information, to be used in the ADS; this was updated in 1998. This subset only contains references which were already in the ADS, thus it is seriously incomplete in referring to articles in the non-astronomical literature. This citation information from ISI spans January 1982-September 1998.

The electronic journals all have machine-readable, web-accessible, reference pages. The ADS points to these with a hyperlink where possible. Several publishers allow us to use these to maintain citation histories; we do this using our reference resolver software. The same software is also used by some publishers to check the validity of their references, pre-publication.

Additionally we use optical character recognition to create reference and citation lists for the historical literature, after it is scanned (Demleiter *et al.* 1999). The different reference parsing processes have handled over 10 million references and added over 6 million parsed references to the ADS citation database.

3.5. COLLABORATION WITH CDS/SIMBAD

The CDS has long maintained several of the most important data services for astronomy (*e.g.* Jung 1971; Jung *et al.* 1973; Genova *et al.* 1998); access to parts of the CDS data via ADS is a key feature of the ADS.

ADS users are able to make joint queries of the ADS bibliographic database and the CDS/SIMBAD bibliographic data base. When SIMBAD contains information on an object which is referred to in a paper whose reference is returned by ADS then ADS also returns a pointer to the SIMBAD data. When a paper has a data table which is kept on-line at the CDS the ADS returns a pointer to it. The CDS-ADS collaboration is at the heart of Urania, a world-wide collaboration of astronomical data providers. More recently the ADS has entered into a collaboration with the NASA Extragalactic Database (NED – Helou & Madore 1988; Madore *et al.* 1992) which is similar to the SIMBAD portion of the CDS-ADS collaboration.

4. User Interface

The ADS services can be accessed through various interfaces (see Eichhorn *et al.* 2000). Some of these interfaces use WWW based forms, others allow direct access to the database and search system through Application Program Interfaces (APIs). This section describes the main WWW interface, as well as the returned results.

4.1. SEARCH FORM

The main query form (Figs. 1, 2, 3) provides access to the different abstract databases. This form is generated on demand by the ADS software. This allows the software to check the user identification through the HTTP (HyperText Transfer Protocol) cookie mechanism (see Eichhorn *et al.* 2000), so that the software can return a customized query form if one has been defined by the user. It also adapts parts of the form according to the capabilities of the user's web browser.

The query form allows the user to specify search terms in different fields. The input parameters in each query field can be combined in different ways, as can the results obtained from the different fields (Fig. 1). The user can specify how the results are combined through settings on the query form (Fig. 3). The combined results can then be filtered according to various criteria (Fig. 2).

The database can be queried for author names, astronomical objects names, title words, and words in the abstract text. References can be selected according to the publication date. The author name, title, and text fields are case insensitive. The object field is case sensitive when the IAU (International Astronomical Union) Circulars object name database is searched, since the IAU object names are case sensitive. In the author and object name fields, the form expects one search term per line since the terms can contain blanks. In the title and text fields line breaks are not significant.

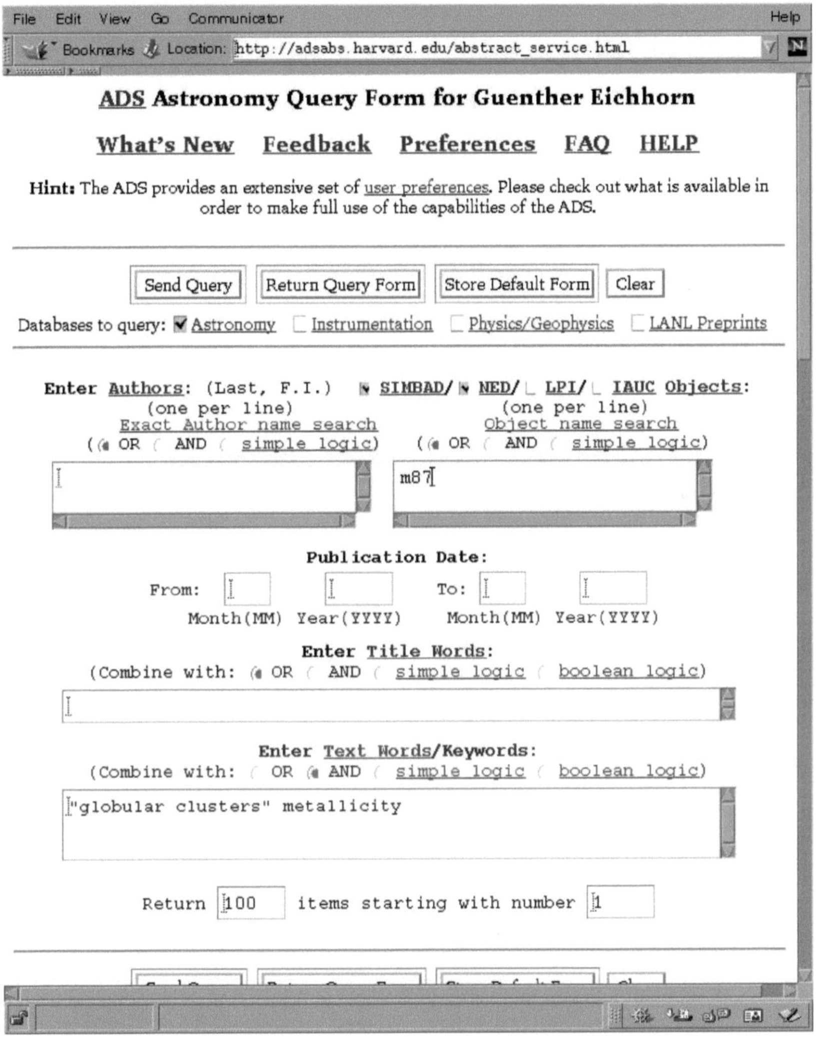

Figure 1. A query to the ADS Abstract Service requesting a listing of papers on the metallicity of M87 globular clusters. SIMBAD, NED, the ADS phrase index, the ADS word index and the ADS synonym list are all queried, the results are combined and the list shown in Fig. 4 is returned.

Object name queries are sent to the requested database(s), which can be external (*e.g.* SIMBAD and NED). The results returned from these external queries are then combined with the other queries to the ADS databases. The resulting combined list is then returned to the user.

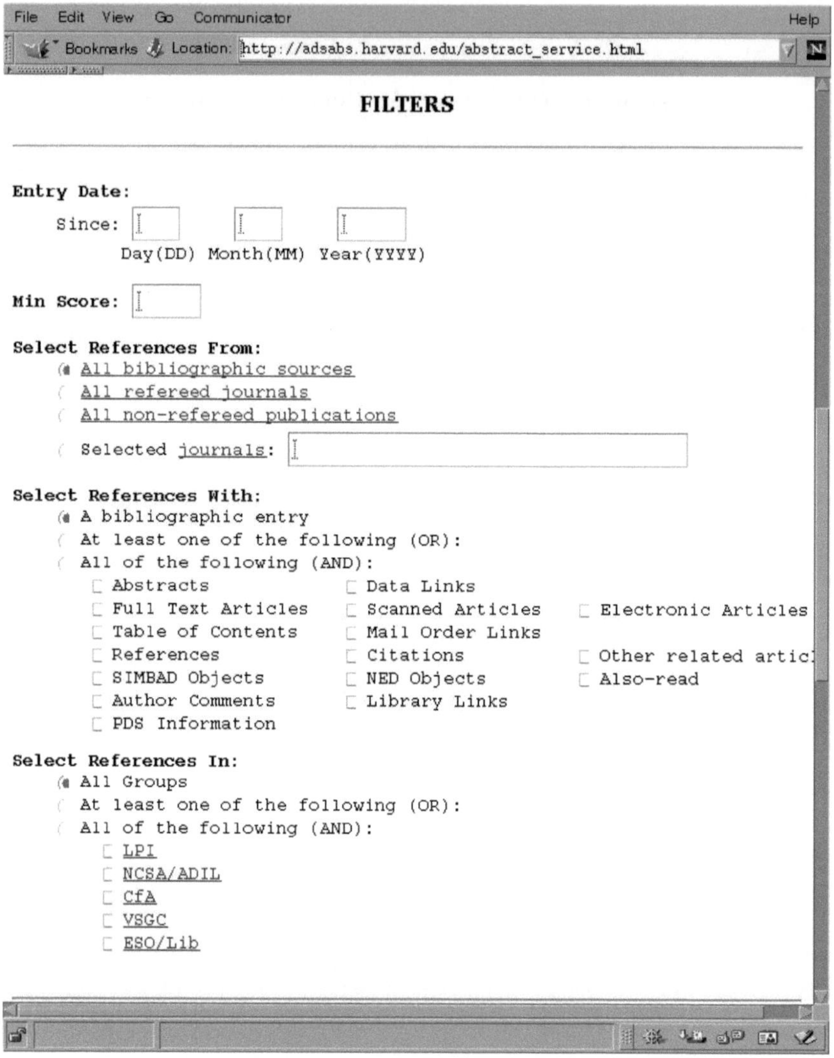

Figure 2. The Filter section of the query form allows selection of references that have specific properties.

4.2. DISPLAY OF SEARCH RESULTS

The ADS system returns different amounts of information about a reference, depending on what the user request was. This section describes the results list format, the format for the full abstract display, and the display for a scanned article.

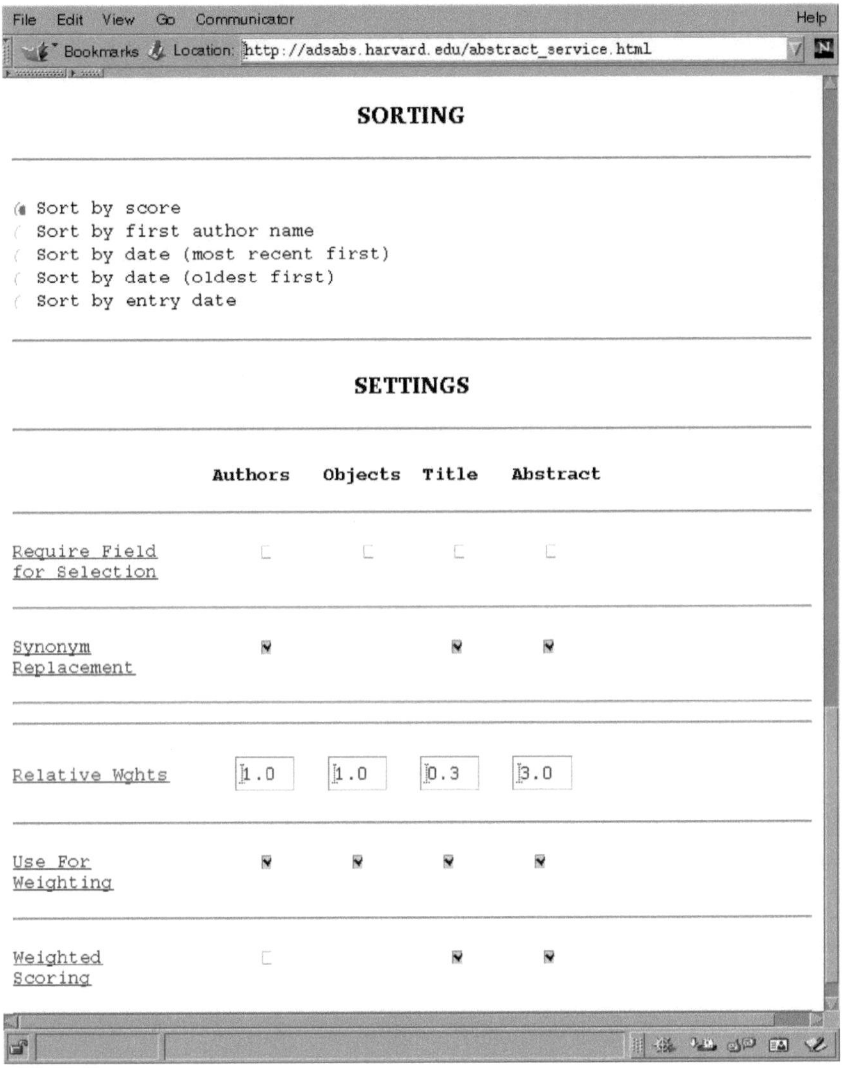

Figure 3. The Settings section of the query form allows the user to customize the search.

4.2.1. *Short Reference Display*

The list of references returned from a query is displayed in a tabular format. The returned references are sorted by score first. For equal scores, the references are sorted by publication date with the latest publications displayed first.

A typical reference display is shown in Fig. 4. The fields in such a reference are shown in Fig. 5. They are as follows:

1. Bibliographic Code: This code identifies the reference uniquely (see Grant *et al.* 2000; Schmitz *et al.* 1995). Two important properties of these codes

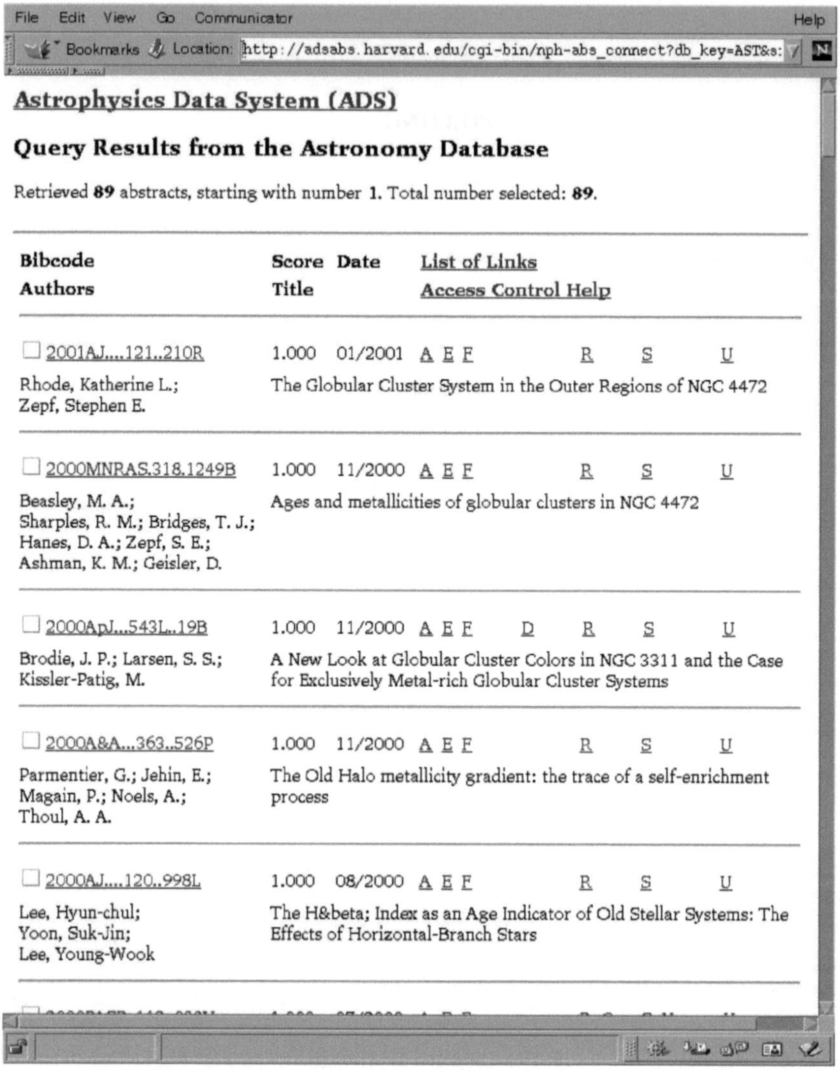

Figure 4. The top of the list ADS returns when the query shown in Fig. 1 is made.

are that they can be generated from a regular journal reference, and that they are human-readable and can be understood and interpreted.

2. Score: The score is determined during the search according to how well each reference fits the query.

3. Date: The publication date of the reference is displayed as mm/yyyy.

4. Links: The links are an extremely important aspect of the ADS. They provide access to information correlated with the article (see Sect. 3.3).

5. Authors: This is the list of authors for the reference. Generally these lists are complete. For some of the older abstracts that we received from

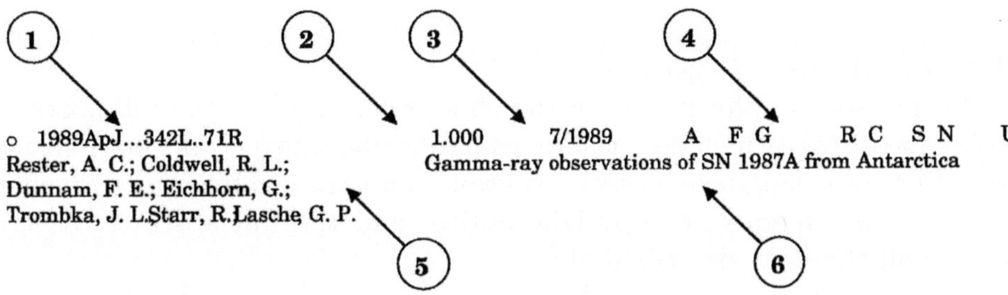

Figure 5. Entries in the list of references returned by an ADS query contain the bibliographic code (1) the matching score (2), the publication date (3), a list of data links (4), the list of authors (5), and the title of the reference (6).

NASA/STI, the author lists were truncated at 5 or 10 authors, but every effort has been made to correct these abbreviated author lists (see Grant *et al.* 2000).
6. Title: The complete title of the reference.

The reference lists are returned as forms if table display is selected (see Eichhorn *et al.* 2000). The user can select some or all of the references from that list to be returned in any one of several formats:
i. HTML format: The HTML (HyperText Markup Language) format is for screen viewing of the formatted record.
ii. Portable Format: This is the format that the ADS uses internally and for exchanging references with other data centers. A description of this format is available on-line[1].
iii. BibTeX format: This is a standard format that is used to build reference lists for TeX (a typesetting language especially suited for mathematical formulas) formatted articles.
iv. ASCII format: This is a straight ASCII text version of the abstract. All formatting is done with spaces, not with tabs.
v. User Specified Format: This allows the user to specify in which format to return the reference. The default format for this selection is the bibitem format from the AASTeX macro package. The user can specify an often used format string in the user preferences (see Eichhorn *et al.* 2000). This format string will then be used as the default in future queries.

[1] `http://adsabs.harvard.edu/abs_doc/abstract_format.html`

The user can select whether to return the selected abstracts to the browser, a printer, a local file for storage, or email it to a specified address.

4.2.2. *Full Abstract Display*

In addition to the information in the short reference list, the full abstract display (see Fig. 6) includes, where available, the journal information, author affiliations, language, objects, keywords, abstract category, comments, origin of the reference, a copyright notice, and the full abstract. It also includes all the links described above.

For abstracts that are displayed as a result of a search, the system will highlight all search terms that are present in the returned abstract. This makes it easy to locate the relevant parts in an abstract. Since the highlighting is somewhat resource intensive, it can be turned off in the user preference settings (see Eichhorn *et al.* 2000).

For convenience, the returned abstract contains links that allow the user to directly retrieve the BibTex or the custom formatted version of the abstract.

The full abstract display also includes a form that provides the capability to use selected information from the reference to build a new query to find similar abstracts. The query feedback mechanism makes in-depth literature searches quick and easy. The user can select which parts of the reference to use for the feedback query (*e.g.* authors, title, keywords, or abstract). The feedback query can either be executed directly, or be returned as a query form for further modification before executing it, for instance to change the publication date range or limit the search to specific journals. This query feedback mechanism is a very powerful means to do exhaustive literature searches and distinguishes the ADS system from most other search systems. A query feedback ranks the database against the record used for the feedback and sorts it according to how relevant each reference is to the search record. The query feedback can be done across databases. For instance a reference from the Astronomy database can be used as query feedback in the preprint database to see the latest work in the field of this article.

If the article for the current reference has been scanned and is available through the ADS Article Service (see below), printing options are available in the abstract display as well. These printing options allow the printing of the article without having to retrieve the article in the viewer first.

4.2.3. *Scanned Article Display*

The article display normally shows the first page (Fig. 7) of an article at the selected resolution and quality (see the section on Preferences in Eichhorn *et al.* 2000). The user can select resolutions of 75, 100, or 150 dots per inch

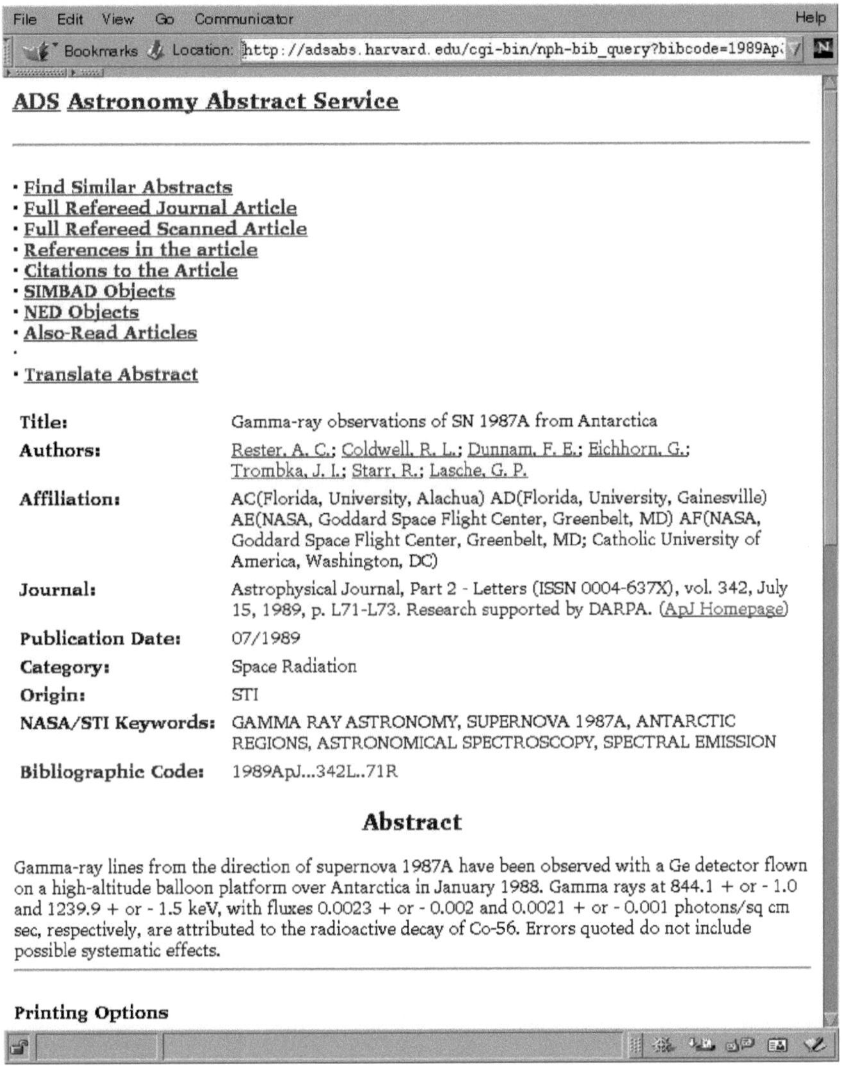
File Edit View Go Communicator Help
Bookmarks Location: http://adsabs.harvard.edu/cgi-bin/nph-bib_query?bibcode=1989Ap

ADS Astronomy Abstract Service

- Find Similar Abstracts
- Full Refereed Journal Article
- Full Refereed Scanned Article
- References in the article
- Citations to the Article
- SIMBAD Objects
- NED Objects
- Also-Read Articles
-
- Translate Abstract

Title:	Gamma-ray observations of SN 1987A from Antarctica
Authors:	Rester, A. C.; Coldwell, R. L.; Dunnam, F. E.; Eichhorn, G.; Trombka, J. I.; Starr, R.; Lasche, G. P.
Affiliation:	AC(Florida, University, Alachua) AD(Florida, University, Gainesville) AE(NASA, Goddard Space Flight Center, Greenbelt, MD) AF(NASA, Goddard Space Flight Center, Greenbelt, MD; Catholic University of America, Washington, DC)
Journal:	Astrophysical Journal, Part 2 - Letters (ISSN 0004-637X), vol. 342, July 15, 1989, p. L71-L73. Research supported by DARPA. (ApJ Homepage)
Publication Date:	07/1989
Category:	Space Radiation
Origin:	STI
NASA/STI Keywords:	GAMMA RAY ASTRONOMY, SUPERNOVA 1987A, ANTARCTIC REGIONS, ASTRONOMICAL SPECTROSCOPY, SPECTRAL EMISSION
Bibliographic Code:	1989ApJ...342L..71R

Abstract

Gamma-ray lines from the direction of supernova 1987A have been observed with a Ge detector flown on a high-altitude balloon platform over Antarctica in January 1988. Gamma rays at 844.1 + or - 1.0 and 1239.9 + or - 1.5 keV, with fluxes 0.0023 + or - 0.002 and 0.0021 + or - 0.001 photons/sq cm sec, respectively, are attributed to the radioactive decay of Co-56. Errors quoted do not include possible systematic effects.

Printing Options

Figure 6. The full abstract display contains (where available) the title, autho r list, journal information, author affiliations, publication date, keywords, the origin of the reference, the bibliographic code, the abstract, object names, abstract category, and a copyright notice.

(dpi) and image qualities of 1, 2, 3, or 4 bits of greyscale per pixel. These gif images are produced on demand from the stored tiff images (see Sect. 3.2 and Grant *et al.* 2000).

Below the page image on the returned page are links to every page of the article individually. This allows the user to directly access any page in the article. Wherever possible, plates that have been printed separately

Astron. Astrophys. 311, 384–392 (1996)

ASTRONOMY AND ASTROPHYSICS

ROSAT observations of BL Lacertae objects

G. Lamer[1], H. Brunner[1,2], and R. Staubert[1]

[1] Institut für Astronomie und Astrophysik, Astronomie, Universität Tübingen, Waldhäuserstr. 64, D-72076 Tübingen, Germany
[2] Astrophysikalisches Institut Potsdam, An der Sternwarte 16, D-14482 Potsdam, Germany

Received 22 August 1995 / Accepted 14 December 1995

Abstract. We present soft X-ray spectra of 74 BL Lacertae objects observed with the PSPC detector on board of the ROSAT satellite. The sample contains all BL Lac objects detected during the pointed observation phase as a target or serendipitously. We have investigated the soft X-ray and broad band spectral properties and discuss the consequences for the X-ray emission processes. For the first time a clear dependence of the X-ray spectral steepness on the radio to X-ray spectral energy distribution is found: α_{rx} and α_x are *correlated* in the X-ray selected (XBL) subsample and *anticorrelated* in the radio selected (RBL) subsample. The objects with intermediate α_{rx} values thus do have the steepest soft X-ray spectra. Simulated PSPC spectra based on a set of simple two component multifrequency spectra are in good agreement with the measurements and suggest a broad range of synchrotron cutoff energies.

We have calculated synchrotron self-Compton beaming factors for a subsample of radio bright objects and find a correlation of the beaming factors δ_{IC} with α_{rx} and α_x. The most extreme RBL objects are very similar to flat spectrum radio quasars in all their broad band and X-ray properties.

Key words: galaxies: active – BL Lacertae objects: general – radio continuuum: galaxies – X-rays: galaxies

1. Introduction

BL Lac objects are active galactic nuclei which by definition do not have strong emission lines, are highly variable, and show strong polarization in the radio to optical emission. Commonly the properties of BL Lacs are explained by the concept that the emission in all spectral bands is dominated by relativistic jets. Relativistic electrons emit synchrotron radiation and may scatter up either synchrotron photons (synchrotron self-Compton

but an increasing number of less radio bright objects were discovered by the identification of X-ray sources (XBLs). Due to the lack of spectral features in the optical no complete optically selected samples exist. The total number of BL Lac objects in the catalogue of Véron-Cetty & Véron (1993) is less than 200. A new catalogue consisting of 233 sources is being published by Padovani & Giommi (1996a). The ongoing search for new BL Lacs from ROSAT sources (Kock et al. 1996, Nass et al. 1996) will significantly increase the number of known objects.

Recently BL Lacs have gained great interest due to the detection of several objects in high energy γ-rays by EGRET on the Compton Gamma Ray Observatory (von Montigny et al. 1995).

There is some evidence that RBLs and XBLs form distinct subclasses as they show a bimodal distribution in the plane of the broad band spectral indices α_{ro} versus α_{ox} (e.g. Giommi et al. 1990). Regardless of the discovery waveband we will use these terms for a distinction of the subclasses based on the spectral energy distribution: RBL is used for *radio* bright objects having $\alpha_{rx} > 0.75$ and XBL for *X-ray* bright objects with $\alpha_{rx} < 0.75$.

There have been several attempts to explain the physical differences of the XBL and RBL objects and their relation to flat spectrum radio quasars (FSRQs). The concept that parts of the continuum emission from radio loud AGN arises from jets of radiogalaxies more or less aligned with the line of sight is widely accepted.

Based on the study of number count relations and luminosity functions several authors (e.g. Padovani & Urry 1990) proposed that BL Lacs are the beamed subpopulation of FR I galaxies with RBLs having higher beaming factors than XBLs. Ghisellini & Maraschi (1989) discussed an "accelerating jet" model with lower bulk Lorentz factors Γ in the X-ray emitting regions resulting in broader beaming cones for the X-ray emission and narrow radio cones. Celotti et al. (1993) developed a

Figure 7. The article display shows a gif image of the selected journal page with the resolution selected in the user preferences.

in the back of the journal volume have been bundled with the articles to which they belong for ease of access. The next part of the displayed document provides access to plates in that volume if the plates for this journal are separate from the articles. Another link retrieves the abstract for this article.

The next part of the page allows the printing of the article. If the browser works with HTTP persistent cookies (see Eichhorn *et al.* 2000), there is just one print button in that section with a selection to print either the whole paper or individual pages. This print button will print the article in the format that the user has specified in the user preferences. If the browser does not handle cookies, several of the more commonly used print options are made available here.

All possible printing options can be accessed through the next link called "More Article Retrieval Options". This page allows the user to select all possible retrieval options. These include:
i. Postscript: Access to two resolutions is provided (200 dpi and 600 dpi). For compatibility with older printers, there is also an option to retrieve Postscript Level 1 files. Postscript is a printer control language developed by Adobe (see Adobe 1990).
ii. PCL (Printer Control Language): This language is for printing on PCL printers such as the HP desk jets and compatibles.
iii. PDF (Portable Document Format): PDF can be viewed with the Adobe Acrobat reader (Adobe 2002). From the Acrobat reader the article can be printed.
iv. TIFF (Tagged Image File Format): The original images can be retrieved for local storage. This would allow further processing like extraction of figures, or Optical Character Recognition (OCR) in order to translate the article into ASCII text.
v. Email retrieval: Articles can be retrieved through email instead of through a WWW browser. MIME (Multipurpose Internet Mail Extension – Vaudreuil 1992) capable email systems should be able to send the retrieved images directly to the printer, to a file, or to a viewer, depending on what retrieval option was selected by the user.

For most of the retrieval options, the data can optionally be compressed before they are sent to the user. Unix compress and GNU gzip are supported compression algorithms.

Instead of displaying the first page of an article together with the other retrieval links, the user has the option (selected through the preferences system, see Eichhorn *et al.* 2000) to display thumbnails of all article pages simultaneously. This allows an overview of the whole article at once. One can easily find specific figures or sections within an article without having to download every page. This should be especially useful for users with slow connections to the Internet. Each thumbnail image ranges in size from only 700 bytes to 3000 bytes, depending on the user selected thumbnail image quality. The rest of this type of article page is the same as for the page-by-page display option.

5. Search Engine

The basic design assumption behind the search engine, and other user interfaces, is that the user is an expert astronomer. This differs from the majority of information retrieval systems, which assume that the user is a librarian. The default behavior of the system is to return more relevant information, rather than just the most relevant information, assuming that the user can easily separate the wheat from the chaff. In the language of information retrieval this is favoring recall over precision.

The search engine uses software written by the ADS project. A first version of the search system that was based on commercial database software turned out to be too slow and have too many limitations. Writing the complete search system ourselves, using various optimization strategies, significantly improved the search speed and provides great flexibility in adapting the search system to user requests for new features. It allows us to easily use new technology as it becomes available, and allows the porting of the ADS server software to several different platforms for various mirror sites.

6. Mirror Sites

Soon after after the inception of the article service in 1995 it became clear that for most ADS users the limiting factor when retrieving data from our computers was bandwidth rather than raw processing power. With the creation of the first mirror site hosted by the CDS in late 1996, users in different parts of the world started being able to select the most convenient database server when using the ADS services, making best use of bandwidth available to them. At the time of this writing, there are ten mirror sites located on four different continents, and more institutions have already expressed interest in hosting additional sites. The administration of the increasing number of mirror sites requires a scalable set of software tools which can be used by the ADS staff to replicate and update the ADS services both in an interactive and in an unsupervised fashion.

The cloning of our databases on remote sites has presented new challenges to the ADS project, imposing additional constraints on the organization and operation of our system. In order to make it possible to replicate a complex database system elsewhere, the database management system and the underlying data sets have to be independent of the local file structure, operating system, and hardware architecture. Additionally, networked services which rely on links with both internal and external web resources (possibly available on different mirror sites) need to be capable of deciding how the links should be created, giving users the option to review and modify the system's linking strategy. Finally, a reliable and efficient mechanism should be in place to allow unsupervised database updates, especially for those applications involving the publication of time-critical data.

The database management software and the search engine used by the ADS bibliographic services have been written to be independent from system-specific attributes to provide maximum flexibility in the choice of hardware and software in use on different mirror sites. We currently support the Sparc/Solaris and x86/Linux servers. Given the current trends in hardware and operating systems, we expect to standardize to GNU/Linux systems in the future.

Hardware independence was made possible by writing portable software that can be either compiled under a standard compiler and environment framework (*e.g.* the GNU programming tools – Loukides & Oram 1996) or interpreted by a standard language (*e.g.* PERL Version 5 – Wall *et al.* 1996). Under this scheme, the software used by the ADS mirrors is first compiled from a common source tree for the different hardware platforms on the main ADS server, and then the appropriate binary distributions are mirrored to the remote sites.

Operating System independence is achieved by using a standard set of public domain tools abiding to well-defined POSIX standards (IEEE 1995). Any additional enhancements to the standard software tools provided by the local operating system is achieved by cloning more advanced software utilities (*e.g.* the GNU shell-utils package) and using them as necessary. Specific operating system settings which control kernel parameters are modified when appropriate to increase system performance and/or compatibility among different operating systems (*e.g.* the parameters controlling access to the system's shared memory). This is usually an operation that needs to be done only once when a new mirror site is configured.

File-system independence is made possible by organizing the data files for a specific database under a single directory tree, and creating configuration files with parameters pointing to the location of these top-level directories. Similarly, host name independence is achieved by storing the host names of ADS servers in a set of configuration files.

The mirroring of the data is completely automated. It is activated either by a program (anytime the database is updated), or manually through a WWW interface (when any part of the data or software have been manually modified).

Currently the ADS is mirrored at 10 sites. Table 2 shows the current mirror sites and their URLs.

Setting up a mirror site is fairly easy. The hosting institution has to provide a server and an Internet connection. Such an abstract mirror site can now run on a Linux PC with 20 Gb of disk space. A partial article mirror site can run on as little as 80 Gb of disk space. If you are interested in having a mirror site, please contact Guenther Eichhorn at `gei@cfa.harvard.edu` for detailed requirements.

7. Query Examples

The ADS answers over 8,000,000 queries per year, covering a wide range of possible query type, from the simplest (and most popular): "give me all the papers written by (some author)," to complex combinations of natural language described subject matter and bibliometric information. Each query is essentially the sum of simultaneous queries (*e.g.* an author query and a title query), where the evidence is combined to give a final relevance ranking (*e.g.* Belkin *et al.* 1995).

Here we show examples of simple, but sophisticated queries, to give an indication of what is possible using the system. A detailed description of available query options is in Eichhorn *et al.* (2000).

Fig. 1 shows how to make the query "what papers are about the metallicity of M87 globular clusters?" This was the first joint query made after the SIMBAD-ADS connection was completed in 1993.

There are 1,914 papers on M87 in SIMBAD, NED, or both; there are 8,546 papers which contain the phrase "globular cluster" in ADS, and there are 33,896 papers in ADS containing "metallicity" or a synonym (abundance is an example of a synonym for metallicity). The result, which comes in a couple of seconds, is a list of just those 89 papers desired.

Five different indices are mixed in this query: the SIMBAD object—bibcode index query on M87 is logically OR'd with the NED object—refcode index query for M87. The ADS phrase index query for "globular cluster" is (following the user's request) logically AND'd with the ADS word index query on metallicity, where metallicity is replaced by its group of synonyms from the ADS astronomy synonym list (this replacement is under user control). If the user requires a perfect match, then the combination of these simultaneous queries yields the list of 89 papers shown in Fig. 4. Before the establishment of the Urania core queries like this were nearly impossible.

Another simple, but very powerful method for making ADS queries is to use the "Find Similar Abstracts" feature. Essentially this is an extension of the ability to make natural language queries, whereby the user can choose one or more abstracts to become the natural language query. This can be especially useful when one wants to read in depth on a subject, but only knows one or two authors or papers in the field. This is a typical situation for many researchers, but especially for students.

As an example, suppose one is interested in the first author's PhD thesis work. Making an author query on "Eichhorn, G" gets a list of his papers,

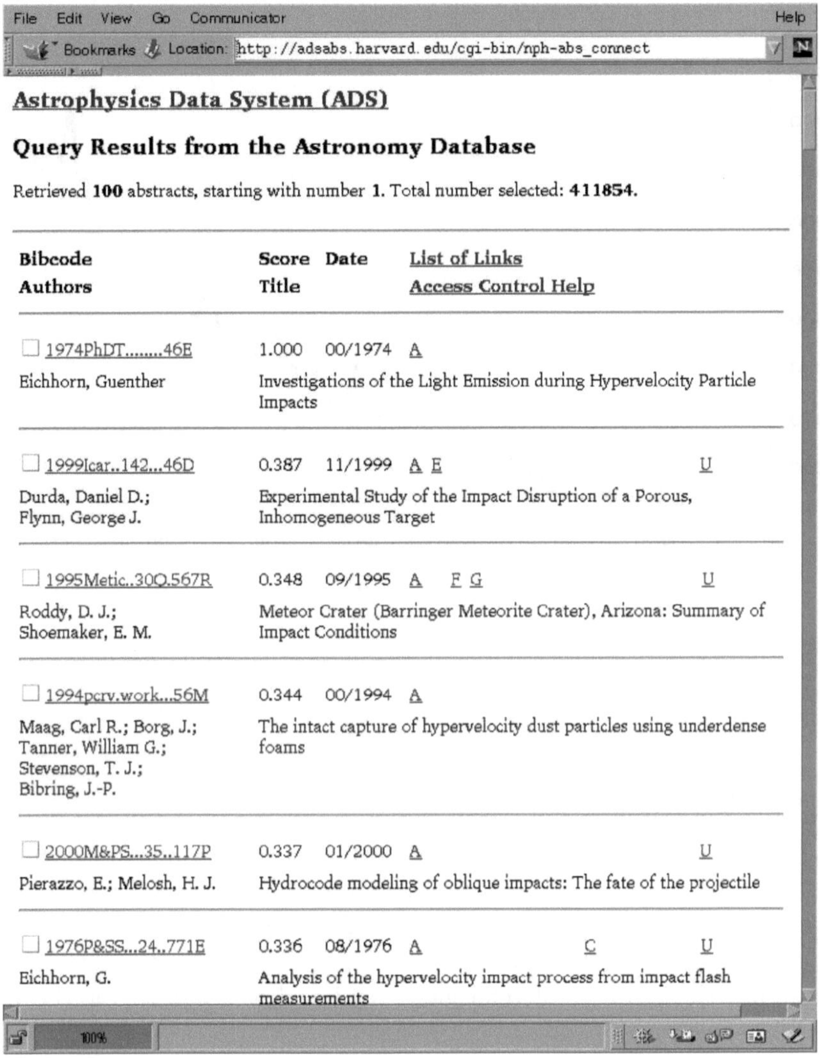
File Edit View Go Communicator Help
Bookmarks Location: http://adsabs.harvard.edu/cgi-bin/nph-abs_connect

Astrophysics Data System (ADS)

Query Results from the Astronomy Database

Retrieved **100** abstracts, starting with number **1**. Total number selected: **411854**.

Bibcode / Authors	Score Date / Title	List of Links / Access Control Help
1974PhDT........46E	1.000 00/1974	A
Eichhorn, Guenther	Investigations of the Light Emission during Hypervelocity Particle Impacts	
1999Icar..142...46D	0.387 11/1999	A E U
Durda, Daniel D.; Flynn, George J.	Experimental Study of the Impact Disruption of a Porous, Inhomogeneous Target	
1995Metic..30Q.567R	0.348 09/1995	A F G U
Roddy, D. J.; Shoemaker, E. M.	Meteor Crater (Barringer Meteorite Crater), Arizona: Summary of Impact Conditions	
1994pcrv.work...56M	0.344 00/1994	A
Maag, Carl R.; Borg, J.; Tanner, William G.; Stevenson, T. J.; Bibring, J.-P.	The intact capture of hypervelocity dust particles using underdense foams	
2000M&PS...35..117P	0.337 01/2000	A U
Pierazzo, E.; Melosh, H. J.	Hydrocode modeling of oblique impacts: The fate of the projectile	
1976P&SS...24..771E	0.336 08/1976	A C U
Eichhorn, G.	Analysis of the hypervelocity impact process from impact flash measurements	

100%

Figure 8. The top of the list of papers returned when Guenther Eichhorn's (1974) thesis is used as the query.

including his thesis. Next one calls up the abstract of the thesis, goes to the bottom of the page, where the "Find Similar Abstracts" feature is found, and clicks the "Send" button. Alternatively, such feedback queries can be executed from the bottom of the first results list. Fig. 8 shows the top of the list returned as a result. These are papers listed in order of similarity to the first author's 1974 thesis; note that the thesis itself is on top with a score of 1.0, as it matches itself perfectly. This list is a detailed subject matter selected custom bibliography.

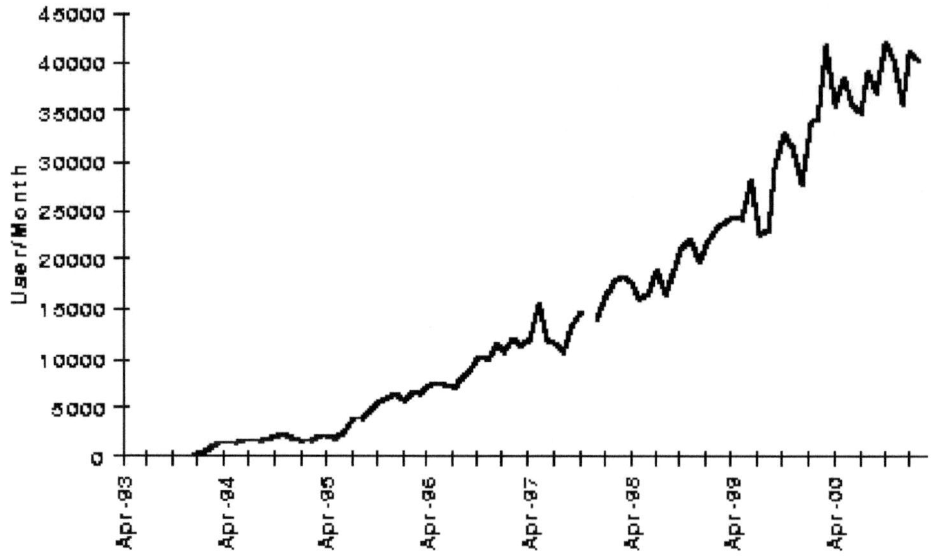

Figure 9. Number of references retrieved per month.

8. Use of the System

The ADS is used by a large majority of professional astronomers world-wide on a daily basis, as well as by many other researchers and non-scientists. This section shows some of the access statistics of the ADS.

The usage of the ADS continuously increased since its start in 1993. Fig. 9 shows the number of references retrieved per month for the life of the ADS.

In September, 2002 ~60,000 users made ~1.1 million queries, and received ~45 million bibliographic references, ~300,000 full text articles and 1 million abstracts, as well as citation histories, links to data, and links to other data centers. Of the 300,000 full-text articles accessed through the ADS ~80% were via pointers to the electronic journals.

ADS users access and print (either to the screen, or to paper) more actual pages than are printed in the press runs of any journal in astronomy. In September, 2002, 1.4 million page images were downloaded from the ADS archive of scanned bitmaps. About 75% of these were sent directly to a printer, 22% were viewed on the computer screen, and 2% were downloaded into files; viewing thumbnail images make up the rest.

The ADS is used 24 hours per day. The distribution of queries throughout the day is shown in Fig. 10. This figure shows the number of queries at our different mirror sites. The usage distribution data are for the time period from 1 November 2000 to 31 March 2001, not the full year, to avoid complications due to different periods where daylight savings time is in effect. The USA distribution is shown in two parts, one for requests coming from US host (SAO-US), the other for requests coming from non-US hosts (SAO-Non-US).

Most of the individual curves show a distinct two-peaked basic shape, with additional smaller peaks in some cases. This distribution of queries over the day shows the usage throughout a workday, with a small minimum during lunch hour. The distribution of accesses to the US site from US hosts is not quite the same, probably because the US covers 3 time zones.

The distribution for Germany and England are very similar, with the English distribution shifted by 1 hour, as is to be expected because of the time difference. The distribution of accesses to the French mirror site is broader than the German or English distribution, but with the peaks in about the same place. The broader distribution is probably due to the fact that the French mirror site is the oldest, and therefore is used by more people world-wide, which tends to wash out the distinct time dependence.

The shape of the accesses to the ADS mirror in France is the same as the shape of the non-US access to the SAO site. This indicates that the large majority of the non-US use on the SAO site is from European users. This non-US usage at the US site is about three times as high as the total usage of the ADS mirror site at the CDS in France. The reason for this is most probably the fact that the connectivity within Europe is sometimes not yet very good. We know that for instance that our users in England and Sweden have better access to the main ADS site in the USA than to our mirror site in France. The same is true for other parts of Europe.

Another reason for the use of the USA site by European users is the fact that our European mirror sites do not yet have the complete set of scanned articles on-line. This forces some users to access the main ADS site in order to retrieve scanned articles.

There is a slight peak in the distribution of queries to the NAO mirror in Japan around 21:00 UTC (Universal Time Coordinated, formerly Greenwich Mean Time). This is probably due to US west coast users using the Japanese mirror site instead of the US site. The access to Japan is frequently very fast and response times from Japan may be better than from SAO during peak traffic times.

The distribution of accesses to China, as expected, is somewhat similar to the distribution of accesses to Japan. The access to Chile peaks about 5 hours after the accesses to Europe as expected. The statistics for accesses

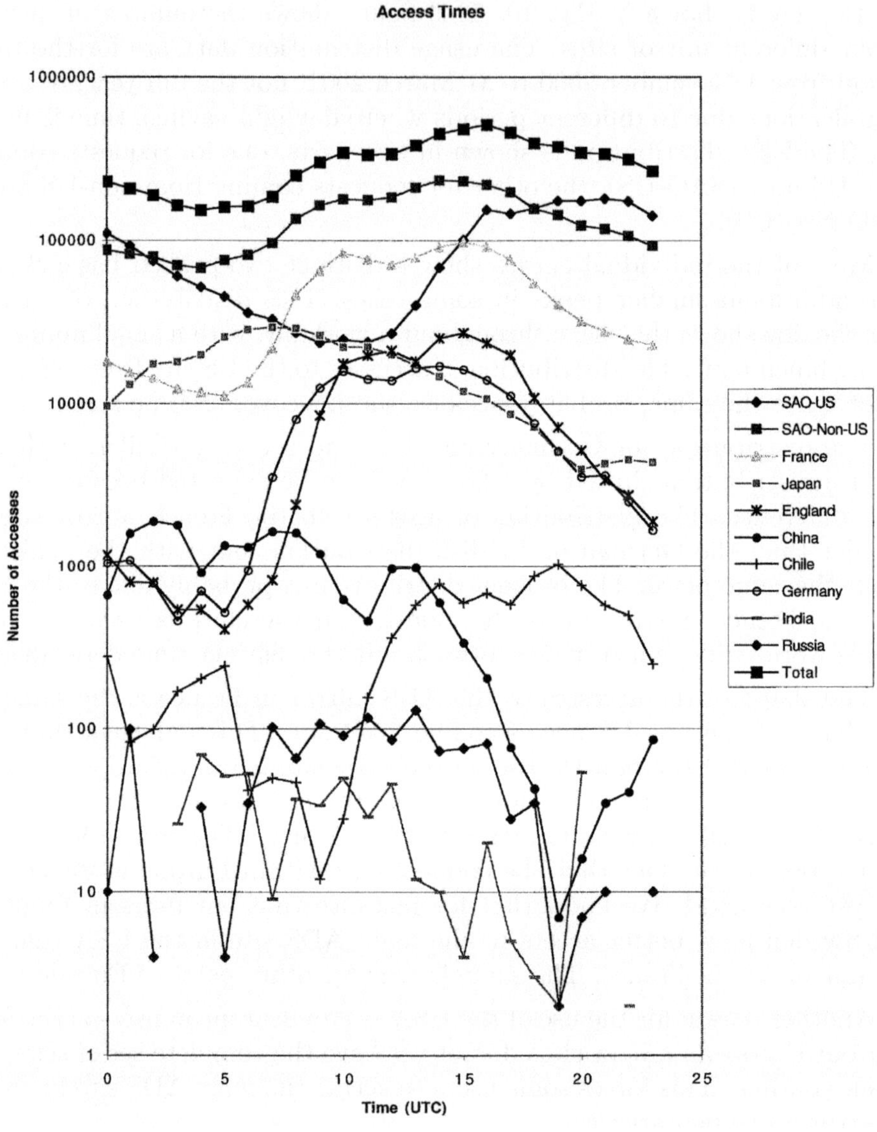

to the mirror sites in India and Russia are not good enough to allow any comparison.

In total, the usage from US hosts is about 1/3 of the total usage of the ADS, 2/3 is from non-US sites, mostly from Europe.

9. Conclusion

The ADS provides free access to most of the astronomical literature. It has profoundly changed the way astronomers do their research. We hope that it will continue to facilitate astronomical research in particular in countries that do not have easy access to libraries with astronomical literature. It should also allow new studies of the historical literature that are so far very difficult or impossible. We welcome any questions and suggestions on how to improve the ADS services. Please contact us at `ads@cfa.harvard.edu` .

Acknowledgment

Funding for this project has been provided by NASA under NASA Grant NCC5-189.

References

1. Accomazzi, A., Eichhorn, G., Grant, C.S., Kurtz, M. J. & Murray, S.S. 2000, The NASA Astrophysics Data System: Architecture, *A&AS* **143**, 85.
2. Adobe 1990, Postscript Language Reference Manual – Second Edition, Addison-Wesley, Reading MA.
3. Adobe 2002, Adobe Acrobat Reader `http://www.adobe.com/prodindex/acrobat/alternate.html` .
4. Belkin, N.J., Kantor, P., Fox, E.A. & Shaw, J.A. 1995, Combining the Evidence of Multiple Query Representations for Information Retrieval, *Information Processing and Management* **31**, 431.
5. Boyce, P.B. 1995, The Electronic ApJ Letters, *American Astronomical Society Meeting*, **187**, 3801.
6. Boyce, P.B. 1996, Journals, Data and Abstracts Make an Integrated Electronic Resource, *American Astronomical Society Meeting* **189**, 603.
7. Corbin, B.G. & Coletti, D.J. 1995, Digitization of Historical Astronomical Literature, *Vistas in Astronomy* **39**, 161.
8. Demleitner, M., Accomazzi, A., Eichhorn, G., Grant, C.S., Kurtz, M.J. & Murray, S.S. 1999, Looking at 3,000,000 References Without Growing Grey Hair, *American Astronomical Society Meeting* **195**, 8209.
9. Egret, D., Wenger, M. & Dubois, P. 1991, The SIMBAD Astronomical Database, in *Databases & On-line Data in Astronomy*, D. Egret & M. Albrecht, Eds, Kluwer Acad. Publ., Dordrecht, p. 79
10. Eichhorn, G. 1994, An Overview of the Astrophysics Data System, *Experimental Astronomy* **5**, 205.
11. Eichhorn, G., Kurtz, M.J., Accomazzi, A., Grant, C. S. & Murray, S.S. 1994, Full Journal Articles in the ADS Astrophysics Science Information and Abstract Service, *American Astronomical Society Meeting* **185**, 4104.
12. Eichhorn, G., Accomazzi, A., Grant, C.S., Kurtz, M.J. & Murray, S.S. 1995, Access to the Astrophysics Science Information and Abstract System, *Vistas in Astronomy* **39**, 217.
13. Eichhorn, G., Murray, S.S., Kurtz, M.J., Accomazzi, A. & Grant, C.S. 1995, The New Astrophysics Data System, in *Astronomical Data Analysis Software and Systems IV*, *ASP Conf. Ser.* **77**, p. 28.
14. Eichhorn, G., Accomazzi, A., Grant, C.S., Kurtz, M.J. & Murray, S.S. 1996, Various Access Methods to the Abstracts in the Astrophysics Data System, in *Astronomical Data Analysis Software and Systems V*, *ASP Conf. Ser.* **101**, p. 569.
15. Eichhorn, G. 1997, The digital library of the Astrophysics Data System, *Astrophys. Space Sci.* **247**, 189.
16. Eichhorn, G., Kurtz, M.J., Accomazzi, A. & Grant, C.S. 1997, Historical Literature in the ADS, *American Astronomical Society Meeting* **191**, 3502.

17. Eichhorn, G., Kurtz, M.J., Accomazzi, A., Grant, C.S. & Murray, S.S. 2000, The NASA Astrophysics Data System: The search engine and its user interface, *A&AS* **143**, 61.
18. Garfield, E. 1979, Citation Indexing: Its Theory and Application in Science, Technology, and Humanities, John Wiley, New York.
19. Genova, F., Bartlett, J.G., Bonnarel, F., Dubois, P., Egret, D., Fernique, P., Jasniewicz, G., Lesteven, S., Ochsenbein, F. & Wenger, M. 1998, The CDS Information Hub, *Astronomical Data Analysis Software and Systems VII ASP Conf. Ser.* **145**, 470.
20. Grant, C.S., Kurtz, M.J. & Eichhorn, G. 1994, The ADS Abstract Service: One Year Old, *American Astronomical Society Meeting* **184**, 2802.
21. Grant, C.S., Eichhorn, G., Accomazzi, A., Kurtz, M. J. & Murray, S.S. 2000, The NASA Astrophysics Data System: Data holdings, *A&AS* **143**, 111.
22. Helou, G. & Madore, B. 1988, A new extragalactic database, in *Astronomy from Large Databases*, Eds. F. Murtagh & A. Heck, *ESO Conf. Proc.* **28**, p. 335.
23. IEEE Computer Society 1995, 1003-1995 IEEE guide to the POSIX Open System Environment, The Institute of Electrical and Electronics Engineers, Inc.
24. Jung, J. 1971, Report on the Strasbourg Stellar Data Center, *Bull. Info. Centre de Données Stellaires* **1**, 2.
25. Jung, J., Bischoff, M. & Ochsenbein, F. 1973, The catalog of stellar identifications, *Bull. Info. Centre de Données Stellaires* **4**, 27.
26. Kurtz, M.J., Eichhorn, G., Accomazzi, A., Grant, C.S. & Murray, S.S. 2000, The NASA Astrophysics Data System: Overview, *A&AS* **143**, 41.
27. Loukides, M. & Oram, A. 1996, Programming With Gnu Software (Nutshell Handbook), O'Reilly & Associates, Inc.
28. Madore, B.F., Helou, G., Corwin, H.G., Jr., Schmitz, M., Wu, X. & Bennett, J. 1992, The NASA/IPAC Extragalactic Database, In *Astronomical Data Analysis Software and Systems I, ASP Conf. Ser.* **25**, p. 47.
29. Murray, S.S., Brugel, E.W., Eichhorn, G., Farris, A., Good, J.C., Kurtz, M.J., Nousek, J.A. & Stoner, J.L. 1992, in *Astronomy from Large Databases II*, Eds. A. Heck & F. Murtagh, p. 387.
30. Schatz, B.R. & Hardin, J.B. 1994, *Science* **265**, 895.
31. Schmitz, M., Helou, G., Dubois, P., Lague, C., Madore, B., Corwin, H. G., Jr. & Lesteven, S. 1995, NED and SIMBAD Conventions for Bibliographic Reference Coding, in *Information & On-line Data in Astronomy* Eds. D. Egret & M.A. Albrecht, Kluwer Acad. Publ., Dordrecht, p. 259.
32. Vaudreuil, G. 1992, MIME: Multi-Media, Multi-Lingual Extensions for RFC 822 Based Electronic Mail, *ConneXions*, p. 36.
33. all, L., Christiansen, T. & Schwartz, R. 1996, Programming PERL, O'Reilly & Associates, Inc., 2nd ed.
34. World Wide Web Consortium 1999, `http://www.w3.org/`

TABLE 1. Link types in the ADS database.

Link	Resource	Description
A	Abstract	Full abstract of the article. These abstracts come from different sources.
C	Citations	A list of articles that cite the current article. This list is not necessarily complete (see 'R' References).
D	On-line Data	Links to on-line data at other data centers.
E	Electronic Article	Links to the on-line version of the article. These on-line versions are in HTML format for viewing on-screen, not for printing.[a]
F	Printable Article	Links to on-line articles in PDF or Postscript format for printing.[a]
G	Gif Images	Links to the images of scanned articles in the ADS Article Service.
I	Author Comments	Links to author supplied additional information (*e.g.* corrections, additional references, links to data),
L	Library Entries	Links to entries in various library on-line systems.
M	Mail Order	Links to on-line document delivery systems at the publisher/owner of the article.
N	NED Objects	Access to lists of objects for the current article in the NED database.
O	Associated Articles	A list of articles that are associated with the current article. These can be errata or other articles in a series.
P	Planetary Data System	Links to datasets at the Planetary Data System.
R	References	A list of articles referred to in the current article. For older articles these lists are not necessarily complete, they contain only references to articles that are in the ADS database. For some articles that are on-line in electronic form, the 'R' link points to the on-line reference list, and therefore the complete list of references in that article.[a]
S	SIMBAD Objects	Access to lists of objects for the current article in the SIMBAD database.
T	Table of Contents	Links to the list of articles in a books or proceedings volume.
U	Also-Read Articles	Links to the list of articles that were read by the same people that read the current article.

[a] There is generally access control at the site that serves these on-line articles

TABLE 2. ADS Mirror Sites.

Country	Mirror Site	URL
USA	Harvard-Smithsonian CfA, Cambridge, MA	http://ads.harvard.edu
France	Centre des Données astronomiques de Strasbourg	http://cdsads.u-strasbg.fr
Japan	National Astronomical Observatory, Tokyo	http://ads.nao.ac.jp
Chile	Pontificia Universidad Católica, Santiago	http://ads.astro.puc.cl
Germany	European Southern Observatory, Garching	http://esoads.eso.org
Great Britain	University of Nottingham, Nottingham	http://ukads.nottingham.ac.uk
China	Beijing Astronomical Observatory, Beijing	http://baoads.bao.ac.cn
India	Inter-University Centre for Astron. and Astroph., Pune	http://ads.iucaa.ernet.in
Russia	Institute of Astr., Russian Acad. of Sci., Moscow	http://ads.inasan.rssi.ru
Brazil	Observatorio Nacional, Rio de Janeiro	http://ads.on.br

FROM EARLY DIRECTORIES
TO CURRENT YELLOW-PAGE SERVICES

A. HECK
Observatoire Astronomique
11, rue de l'Université
F-67000 Strasbourg, France
heck@astro.u-strasbg.fr

Abstract. This chapter reviews major astronomical directories of the 20th century as well as the corresponding online yellow-page services. Some of the online resources were already operational before the advent of the World-Wide Web which they used subsequently with a diversification of the services offered. While mere collections of commented URLs might loose their utility because of the availability on the web of increasingly better search engines, these facilities do not make obsolete well-targeted, value-added, authenticated, homogeneously presented and continually updated resources (such as the *Star*s Family* and the *StarPages* ones). Additional comments are offered on the rewarding schemes for such activities as well as on copyright protection issues.

1. Introduction

It might be difficult to believe it today, but astronomical directories were already available at the beginning of the 20th century. A new generation appeared in the late seventies and another one with the World-Wide Web (WWW) in the early nineties. These resources included sometimes not only typical data on astronomy-related organizations, but also entries of general interest. These, in turn, reflected how diverse astronomy had become and how diverse too were (and still are) the fields connected to our activities.

With the apparition of networks, resources were put on line, even before the advent of the WWW, and were of course transferred onto the web when it became operational. Typical WWW-centered yellow-page services

A. Heck (ed.), Information Handling in Astronomy – Historical Vistas, 183-205.

became quickly available, but their utility is questioned today with the excellent efficiency of search engines.

This paper will review major astronomy-related directories and information resources, first on paper, then electronic, in the course of the 20th century. It will conclude with a discussion on the future of such resources and on copyright protection issues.

2. First Directories

The first astronomical directory of last century has been produced at the Royal Observatory of Belgium (ROB) by P. Stroobant *et al.* (1907) under the title *"Les observatoires astronomiques et les astronomes"* [1].

In the foreword of the book, P. Lecointe, Scientific Director of ROB's Astronomical Service, tells the genesis of that compilation. Back in 1902, a series of internal lectures by ROB scientists had been launched in order to foster studies in common and to homogenize as much as possible research activities carried out at ROB. To remain up to date with the latest astronomical investigations, a group of people (that took the name of *Comité de bibliographie et d'études astronomiques* [2]) started compiling journals as well as publications of observatories and learned societies. They built up a bibliographical catalogue with the purpose to complement, whenever needed, the *Astronomische Jahresbericht* and the *International Catalogue of Scientific Literature (Astronomy).*

The compilation by Stroobant *et al.* (1907) was one of the first by-products of that work. Lecointe stressed the practical and most useful character of the work "expressing remarkable cooperation spirit and scientific solidarity".

The presentation of each entry in French is literary in the sense that each type of data is fully described. For an observatory, geographical coordinates are given with the reference of the source, as well as a list of personnel (with the corresponding titles and positions), a few historical notes and a description of the instruments and activities. The observatories are listed alphabetically on the location name. The compilation includes a few private facilities.

Astronomical societies are also listed with details on their foundation, aims, activities, publications and constitution of the board. In addition, a few journals are mentioned with indications on their foundation, editors, publication frequencies and prices as well as on their contents and volume status. Finally, indices of names of locations are given together with a few pages where societies and journals are sorted by countries. The compilation

[1] (The) Astronomical Observatories and (the) Astronomers.
[2] Committee of Bibliography and Astronomical Studies.

included also a map showing the distribution of institutions at that time. It has been reproduced here in Fig. 1.

A second edition (Stroobant *et al.* 1931) and a supplement (Stroobant *et al.* 1936) appeared much later with the blessing of the International Astronomical Union (IAU). At its 1952 General Assembly in Rome, the IAU expressed the wish to see the complilation published again. This was done by F. Rigaux in 1959 with a supplement two years later (Rigaux 1959 & 1961). Those directories were again published in French.

From an IAU Commission 5 report (Pecker 1979), it did not seem that further updates or development of the compilations carried out at ORB were intended.

3. A New Generation

3.1. IDAAS, IDPAI, AND THEIR MERGING

We were led to compile astronomical directories quite differently from Stroobant and his collaborators.

After putting together an astronomical photographic atlas (Heck & Manfroid 1977), we needed to make sure that the publisher had all necessary information to sell our masterpiece to the astronomical community and possibly well beyond. We then started gathering lists of essentially amateur organizations around the world. They were indeed the prime target for marketing the volume. Lists of journals were also set up for taking advantage of the reviewing system.

It was subsequently realized that such lists had their own intrinsic interest. They would improve or make easier national and international contacts in amateur astronomy. They would also provide professional astronomers with addresses of groups they could approach for *e.g.* complementary observations. They had also a historical value since they were providing snapshots of the amateur world in those years.

Thus came to light the 1978, 1979, 1981, 1982 and 1984 versions of the IDAAS directory under its original title *"International Directory of Amateur Astronomical Societies"*. A few specific computer printouts were also produced at request between the 1978 and 1979 editions. Later on, IDAAS' meaning was changed to *"International Directory of Astronomical Associations and Societies"* since more and more mixed amateur-professional societies were included and the word *amateur* was not anymore quite appropriate in the title. Three more editions (1986, 1988 and 1990) were produced.

Along the years, an increasing number of professional institutions showed an interest in the successive IDAAS editions. Therefore a list of institutions was also compiled for advertising IDAAS and, since that list

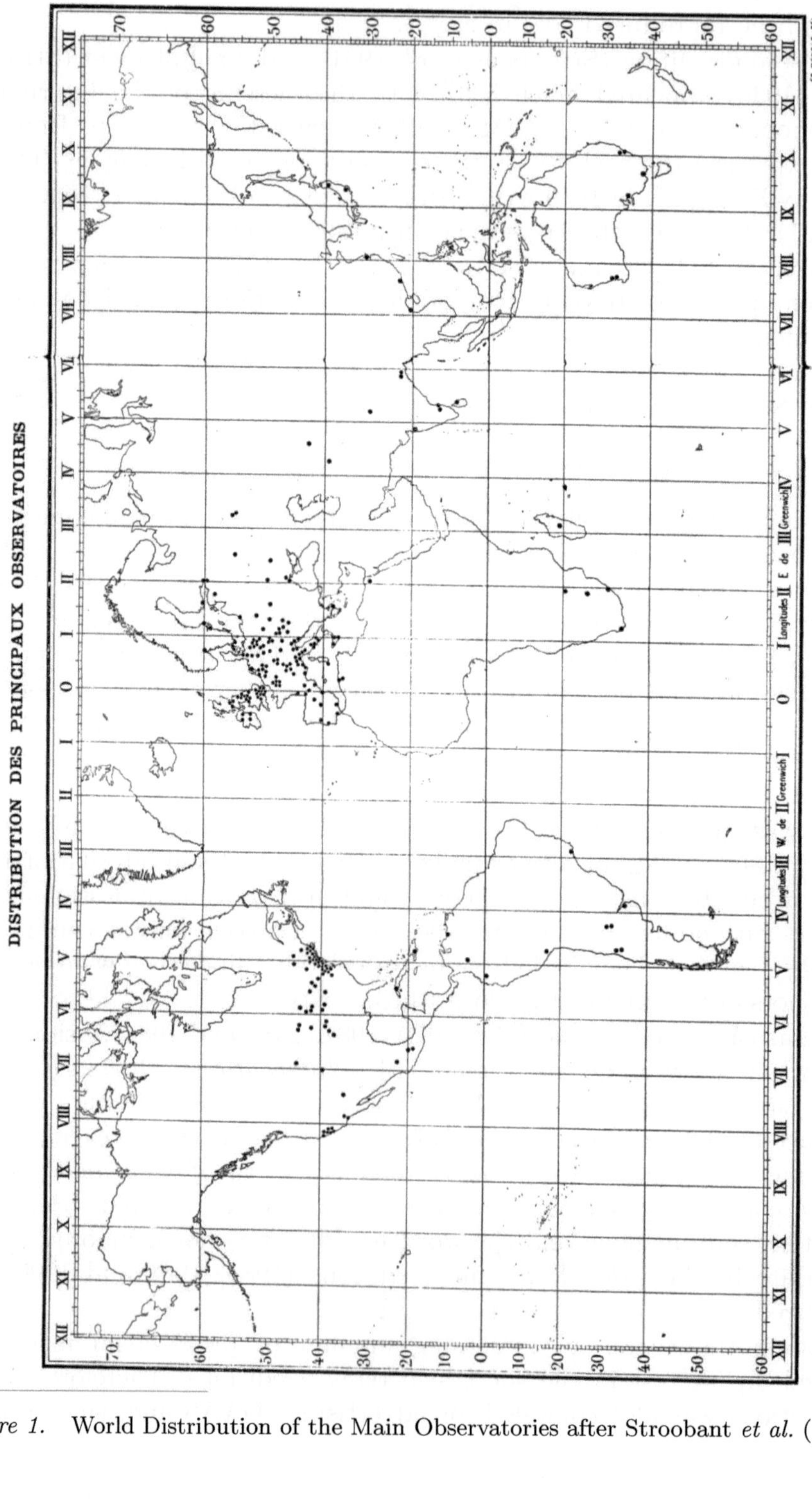

Figure 1. World Distribution of the Main Observatories after Stroobant *et al.* (1907).

was now existing, why not to publish it too? Hence came to light the other directory IDPAI, standing for *"International Directory of Professional astronomical Institutions"*. Three editions have been produced: 1987, 1989 and 1990.

From the start, these directories received an enthusiastic welcome and, along the years, numerous letters of support encouraged us to continue the work and to broaden the scope of the compilations. Therefore the last editions were bearing the subtitle *"together with items of general interest"*.

As the overlap between the two directories grew larger, it was decided to merge them, for the year 1991, into a single one entitled *"Astronomy, Space Sciences and Related Organizations of the World" (ASpScROW)*.

Before going on with the history, it is probably appropriate to say now a few words on the contents of those directories.

3.2. CONTENTS

Those directories were gathering together all practical data available on associations, societies, scientific committees, agencies, companies, institutions, observatories, universities, etc., more generally organizations, involved in astronomy and related space sciences.

Many other entries were also included such as academies, advisory and expert committees, bibliographical services, data and documentation centers, dealers, distributors, funding agencies and organizations, journals, manufacturers, meteorological services, museums, norms and standards offices, planetariums, private consultants, public observatories, publishers, software producers and distributors, and so on.

Other fields such as aeronautics, aeronomy, astronautics, atmospheric sciences, chemistry,communications, computer sciences, data processing, education, electronics, energetics, engineering, environment, geodesy, geophysics, information handling, management, mathematics, meteorology, optics, physics, remote sensing, and so on, were also covered when justified.

The information was given in an uncoded way for easy and direct use. For each entry, all practical data available were listed: city, postal and electronic-mail addresses; URLs; telephone and telefax numbers; foundation years; numbers of members or staff; main activities; titles, frequencies, ISS-Numbers and circulations of periodicals produced; names and geographical coordinates of observing sites; names of planetariums; awards, prizes or distinctions granted; and so on. US FTS numbers were also included in the first versions. Electronic and web addresses were introduced later on (and telex numbers were removed).

The entries were listed alphabetically in each country. At the end of the volumes, an exhaustive index gave a breakdown not only by different designations and acronyms, but also by location and major terms in names. Subindices of academies, awards, bibliographical services, data centers, dealers and distributors, funding organizations, IAU-adhering organizations, ISS-Numbers, journals, manufacturers, meteorological offices, observatories, planetariums, publishers, software producers, etc., were also provided as well as statistics on the contents (numbers of entries per country, memberships, years of foundation) and a list of telephone, telefax and telex national codes.

Table 1 gives a brief description of the successive IDAAS, IDPAI and ASpScROW editions[3]. From 1986 onwards, the directories were distributed as Special Publications of Strasbourg astronomical Data Center (CDS). The main language of publication was English (the first editions had bilingual forewords) while the data were at some stage collected in six languages (English, French, Spanish, German, Italian, Portuguese[4]), than reduced to two (English and French) for simple practical reasons.

A list of acronyms was also included in the 1990 editions of IDAAS and IDPAI, but it had become so voluminous that it became more appropriate to provide it as a separate, nevertheless complementary, publication as we shall see in the following section.

TABLE 1. Statistics from Successive Editions of the Directories

Directory	Number of countries	Number of entries	Number of pages
IDAAS 1978	27	≈600	112
IDAAS 1979	56	>1200	290
IDAAS 1981	51	>1100	304
IDAAS 1982	54	>1200	308
IDAAS 1984	56	>1200	282
IDAAS 1986	58	>1100	270
IDAAS 1988	64	>1700	522
IDAAS 1990	90	>3200	724
IDPAI 1987	68	>1500	280
IDPAI 1989	80	>2700	498
IDPAI 1990	90	>3500	666
ASpScROW 1991 (two volumes)	95	≈5200	1192

[3]See also `http://vizier.u-strasbg.fr/~heck/ahdir.htm` for full references on all those directories. The first ones were produced with the collaboration of J. Manfroid.

[4]And occasionally in Chinese and in Russian ...

4. The Star*s Family and the StarPages

4.1. THE FIRST STAR*S

Things changed dramatically at the beginning of the 1990s. Networks were all over, e-mail was well into its invasion, information flow was in an ever higher gear. There was an increasing demand, from both individuals and organizations, for the kind of societal data we were providing – a demand for a dynamic usage that would be satisfied by providing online access to the data.

Everything had also to go faster too, from the collection-verification-update of data to their shaping and their provision online. From an annual (or so) exercise during the previous fifteen years, it became a continuous, increasingly time-consuming process with an ever heavier pressure.

Through its European Space Information System (ESIS) group, the European Space Agency (ESA) became the first institution to make our data on organizations accessible on line as a database called *StarWays* (Heck *et al.* 1992). The European Southern Observatory (ESO) would follow with the online databases *StarGates* (Albrecht & Heck 1993, 1994b) for organizations and *StarWords* (Albrecht & Heck 1993, 1994a) for the list – by now a dictionary – of acronyms and abbreviations mentioned in the previous section.

The *Star*s Family* was born. This was announced in the CDS Information Bulletin (Heck 1992) where the new names were introduced for the paper versions:

- *StarGuides* for the directory of organizations (see also Heck 1993a);
- *StarBriefs* for the dictionary of abbreviations, acronyms, etc. (see also Heck 1993b).

Two more products were also announced:

- *StarLabels*: sets of mailing stickers bearing addresses, essentially delivered at production cost to publishers, manufacturers, and conference organizers;
- *StarSets*: subsets of data occasionally provided under some conditions of requirements and usage (see in Sect. 7.2 the discussion regarding the protection of data).

Tables 2 and 3 give a description of the successive editions of *StarGuides* and *StarBriefs*. All of them were distributed by CDS, except the 2001 ones published by Kluwer Academic Publishers[5]. Decreasing numbers of pages indicate transitions to more compacted printing.

[5]See for instance the web pages `http://vizier.u-strasbg.fr/~heck/ahdir.htm` and `http://vizier.u-strasbg.fr/~heck/ahcata.htm` for full references on the successive editions of *StarGuides* and *StarBriefs* respectively.

TABLE 2. Statistics from Successive Editions of the Directory *StarGuides*

Year of publication	Number of countries	Number of entries	Number of pages
1993	96	≈5200	1184
1994	96	≈5200	888
1995	97	≈5300	822
1996	97	≈5500	924
1997	98	≈5800	964
1998	100	≈6100	1030
1999	101	≈6100	1132
2000	101	≈6200	1148
2001	100	≈6200	1238

TABLE 3. Statistics from Successive Editions of the Dictionary *StarBriefs*

Year of publication	Number of entries	Number of pages
1992	≈50,000	808
1993	>60,000	812
1994	≈70,000	824
1995	≈80,000	750
1996	≈100,000	912
2000	≈140,000	808
2001	>140,000	822

4.2. THE STARPAGES

The World-Wide Web (see *e.g.* Berners-Lee *et al.* 1992 & 1994) knocked at our door with an immediate success: astronomy users and servers were, by the end of 1993, the largest group on the web (Hardin 1993).

CDS in turn decided to make available, through its WWW server, the resources it was distributing on paper (Heck *et al.* 1994), giving birth to the *StarPages*, the web members of the *Star*s Family*:

• the directory of organizations *StarGuides* had its online counterpart named *StarWorlds* [6];

• the dictionary of acronyms, abbreviations, etc., *StarBriefs* had its online counterpart named *StarBits* [7].

[6] `http://vizier.u-strasbg.fr/starworlds.html`
[7] `http://vizier.u-strasbg.fr/starbits.html`

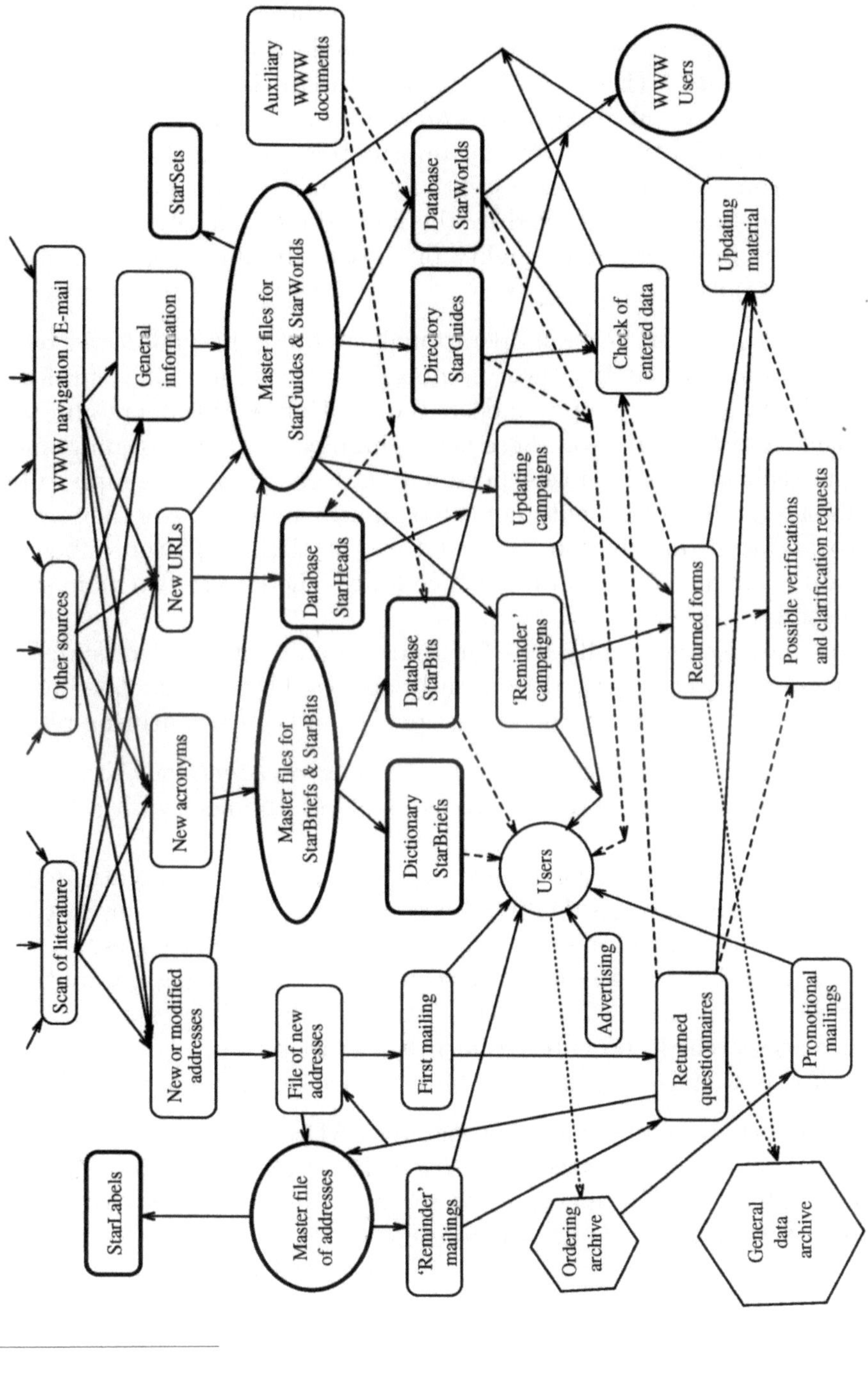

Figure 2. General working scheme for the *Star*s Family* and the *StarPages*.

As astronomers and related scientists started to develop personal homepages on the web, a third database, gathering together the corresponding URLs[8], was set up and made available through the CDS WWW server: *StarHeads* [9] (Heck 1995). That database is purely web-oriented, while the other resources such as *StarGuides* and *StarWorlds* offer much more, as explained above, with an information fully mastered. The data retrieved from *StarHeads* are essentially no more than pointers towards personal homepages maintained by the individual themselves or by the institutional webmasters. The quality of those individual homepages might be very unequal.

When the WWW became operational, many individuals had not realized the enormous visibility those homepages were going to have professionally: real windows on people and on their activities. Students and young scientists – quick to jump on those new "gimmicks" – were not unfrequently filling in their pages with jokes, personal matters, when not simply rubbish of dubious quality often without astronomical interest. Another problem had to be faced quickly too: the volatility of URLs and the difficulty to maintain a valuable compilation of such links. There were two main sources for volatility:

- people moving (especially true for students and young scientists);
- server names were changed frequently and URLs were redefined as webmasters were becoming more "professional" – a typical maturing phase.

A few rules were quickly set up for keeping up *StarHeads'* standards:

- entries would be restricted to professional astronomers with a PhD and above;
- homepages would be included only after being checked, authenticated, and sometimes refereed; occasionally advices would be given for improvement;
- tips would be issued for setting up professionally useful pages;
- a procedure, including systematic scanning of the database, would be set up for checking periodically that the links were alive; if necessary, new links would be researched and entered into the database;
- individuals would systematically be informed of their page's inclusion.

Such efforts were rewarded. For instance, ADS[10] (see *e.g.* Eichhorn *et al.* 2003) decided to point towards *StarHeads.* Requests are continually coming in from individuals for inclusion or for updating the links. As statistics reveal it, that database is in heavy usage.

[8]Unified Resource Locators – the electronic addresses of web pages.
[9]`http://vizier.u-strasbg.fr/starheads.html`
[10]`http://adsabs.harvard.edu/`

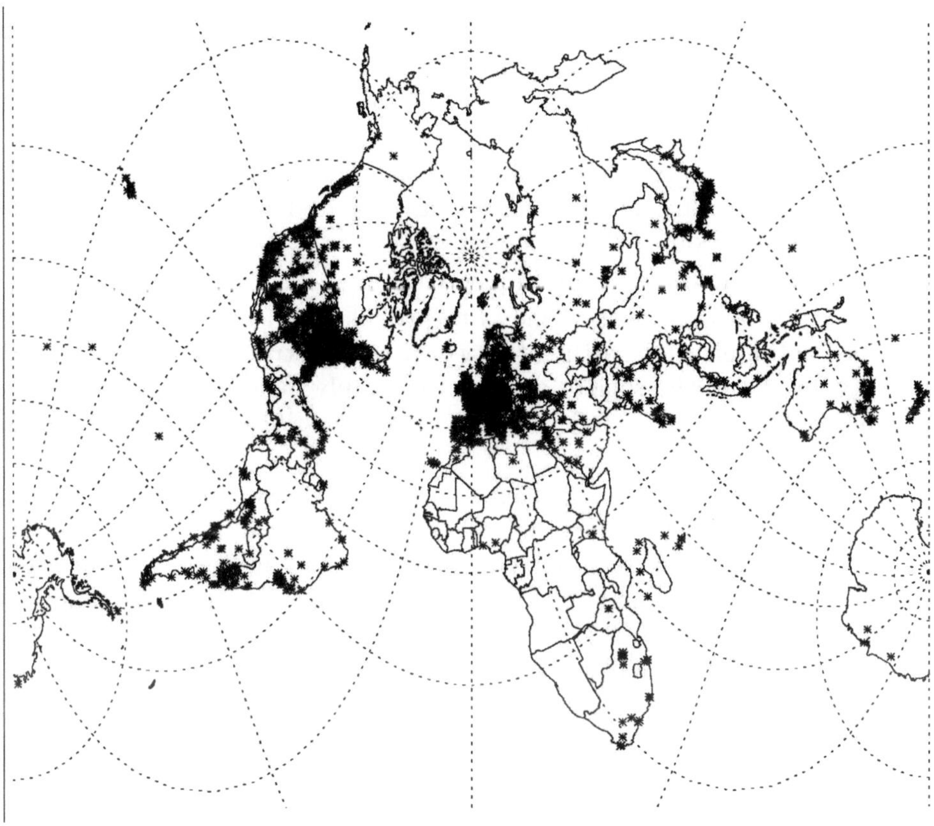

Figure 3. Planet 'Astronomy': World (Heck 2000a).

4.3. CURRENT SITUATION

The *Star*s Family* resources offer features such as a quarter-of-century experience in compilations, a resulting excellent exhaustivity of entries (including also many organizations not yet on the WWW), an homogeneous coverage and presentation of all practical data, as well as a permanent updating and quality checking scheme (see Fig. 2) including authentication of data originators, and so on.

The *StarPages* are currently giving access to the largest amount of WWW links (more than 12,000) available in a set of astronomy/space resources and to a unique astronomy/space-oriented dictionary of acronyms and abbreviations.

The online directory *StarWorlds* of astronomy, space and broadly related organizations of the world gathers currently together about 6,500 entries from 100 countries, offering more than 6,500 WWW links.

The database *StarHeads* of personal web pages of essentially astronomers and space scientists includes currently about 6,000 entries and is also pointed at by services such as the NASA Astrophysics Data System (ADS – Eichhorn *et al.* 2003).

The dictionary *StarBits* of abbreviations, acronyms, contractions, and symbols in astronomy, space sciences, and related fields offers currently explanations for about 200,000 entries.

Updated paper versions are in preparation.

The *Star*s Family* bottomline is to provide services for the benefit of a better communication within the world-wide astronomical community, and between it and the society at large, resolutely putting the emphasis on the quality, on the homogeneity and on the exhaustivity of the information delivered.

Working relationships have been established with various organizations such as (by alphabetical order) the Astronomical League (AL), the Committee for the Scientific Investigations of Claims of the Paranormal (CSICOP), the International Astronomical Union (IAU), the International Organization for Standardization (ISO), NASA's Astrophysics Data System (ADS), and the World Meteorological Organization (WMO).

4.4. SOCIOLOGICAL STUDIES FROM THE STAR*S DATA

Geographical distributions, ages and sizes of astronomy-related organizations have been investigated from comprehensive and up-to-date samples extracted from the master files for *StarGuides/StarWorlds* (for a synthesis, see Heck 2000a). Results for professional institutions, associations, planetariums, and public observatories have also been presented, as well as specific distributions for astronomy-related publishers and commercial-software producers.

The highly uneven general pattern displayed by geographical distributions (see Fig. 3-6) is still very much the same as it was at the beginning of the 20th century (Fig. 1), even if the densities are higher – another illustration of the well-known socio-economic effect of self-reinforcement. Other geographical peculiarities (local concentrations, national cultures and policies, electronic astronomy, ...) have been discussed, as well as the uneasy separation between amateur and professional astronomers in associations.

Some events had a clear impact on the rate of foundation of astronomy-related organizations, such as the World Wars I and II, the beginning of space exploration, the landing of man on the Moon, the end of the Cold

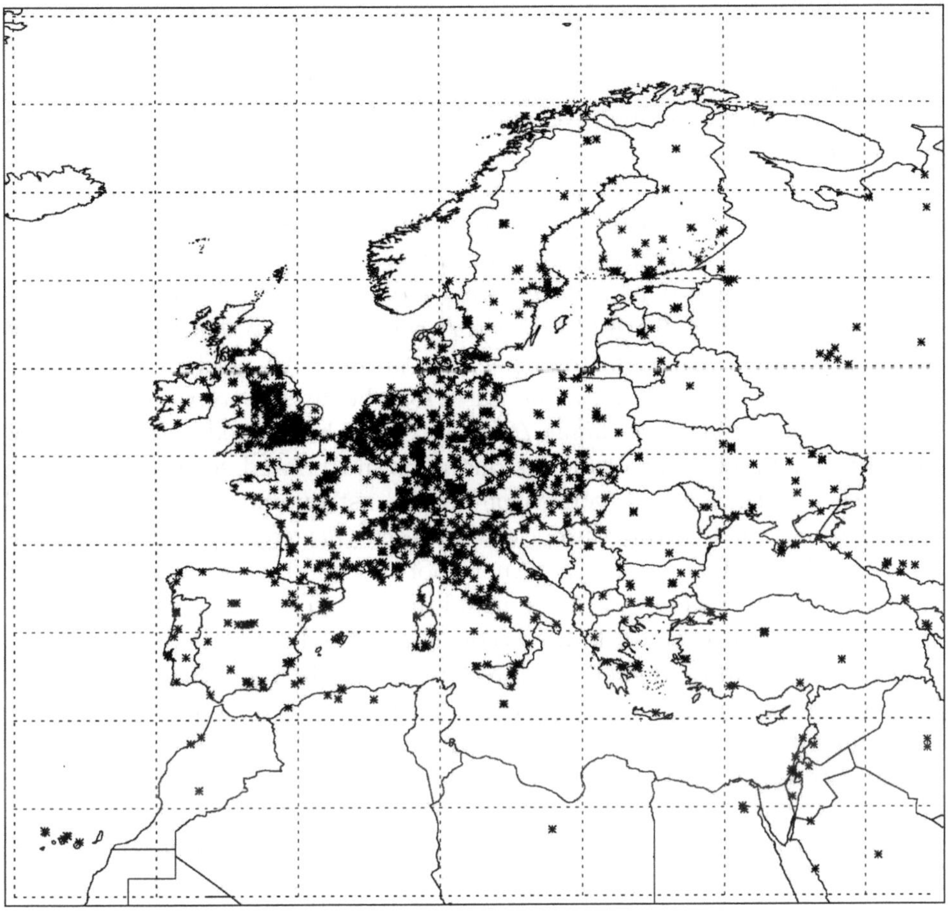

Figure 4. Planet 'Astronomy': Western Europe (Heck 2000a).

War, spectacular comets, and so on. However, as detailed in Heck (2000a), not all of them affected in the same way Western Europe and North America, nor the various types of organizations.

If the size of the vast majority of astronomy-related organizations is relatively small, there are however some differences between Western Europe and North America. See Heck (2000a) and the references quoted therein for details.

5. Other Directories and Resources

5.1. MANY MORE RESOURCES

There are quite a number of other directories in astronomy and space sciences and it would be impossible to quote all of them, especially because the web has allowed them to multiply countlessly – often in an ephemeral way. Some directories on the web have not been maintained at all since long. We shall therefore restrict the following to a few significant resources, keeping in mind that their updatedness and the degree of details they provide is quite unequal.

First of all, the *International Astronomical Union (IAU)* is maintaining an online directory[11] of its worldwide professional membership.

National professional associations publish often membership directories. For instance, the members of *American Astronomical Society (AAS)* receive each year a very useful printed directory where the AAS members are listed alphabetically, then grouped per institutions. An electronic version[12] is also available. This is also the case for the French *Société Française d'Astronomie et d'Astrophysique (SF2A)* [13] that has issued a directory, not only of its members, but of all astronomers and related scientists working in French institutions, as well as of French astronomers working abroad in institutions such as the *European Southern Observatory (ESO)* and the *European Space Agency (ESA)*. An important distinctive feature of that directory is to provide, beyond the usual individual data, a scientific profile for each entry and then an index according to the keywords listed.

Other large national societies make their membership directories available only on paper, sometimes because national law regulates strictly the availability of data (especially electronic ones) on individuals. We would suggest interested parties to check first the web pages of the societies or the organizations, for instance through the online resource *StarWorlds* [14] discussed earlier. If an online membership directory is available, it is generally directly reachable from the main homepage. Otherwise details can be found online for contacting the societies and obtaining a copy of the latest directory on paper.

As a matter of interest, the *Astronomische Gesellschaft (AG)* has published in 1996 an interesting booklet (Klare 1995) listing members with ID pictures plus a few lines on their career and profile.

The US Naval Observatory has also produced a list of active professional observatories (Lukac & Miller 2000) based in part on our own resources: "... StarGuides (and its preceding editions) ... was used extensively ... If one is working with observatories in general, this set should be on the shelf for reference".

The overall availability of e-mail has led to numerous compilations of electronic addresses that people have made available to each other over

[11]`http://www.iau.org/cgi-iau/iau_mem.cgi`
[12]`https://members.aas.org/directory/directory.cfm`
[13]`http://www.cesr.fr/sf2a/`
[14]`http://vizier.u-strasbg.fr/starworlds.html`

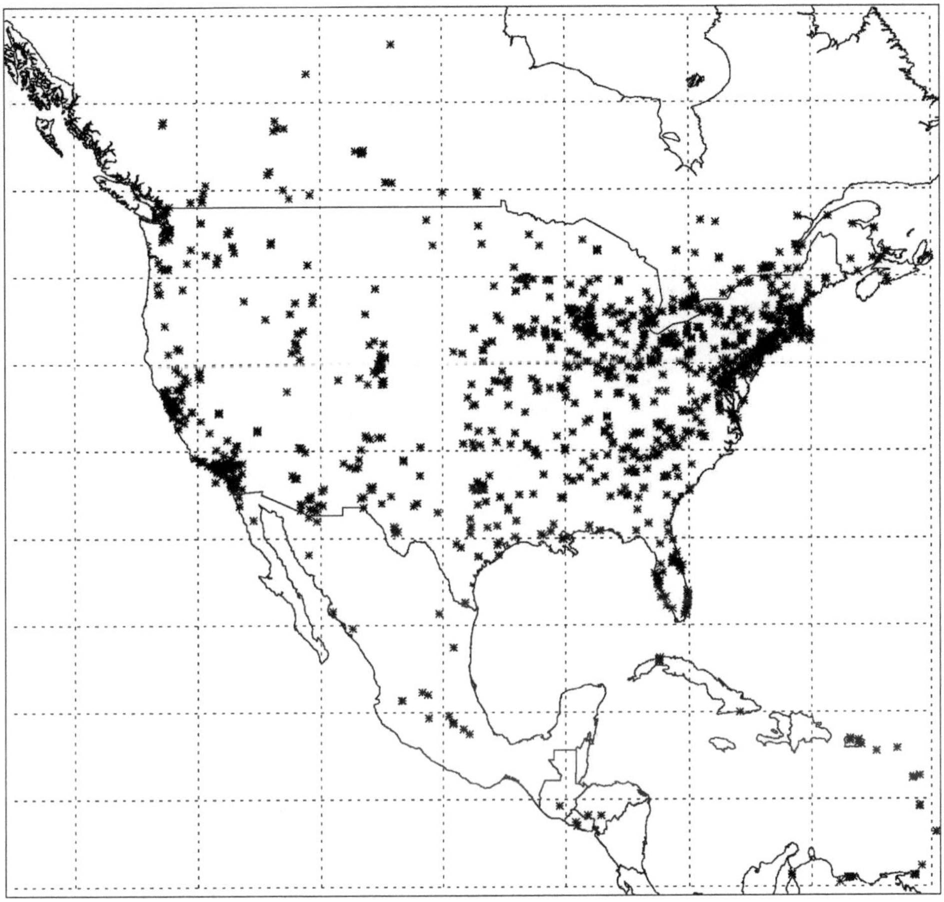

Figure 5. Planet 'Astronomy': North America (Heck 2000a).

the networks. The situation in this field is however evolving quite rapidly and many of these lists have become quickly obsolete. Additionally, the nuisance of spamming and unsollicited e-mailing has led people to become more careful is letting their e-address readily available and in distributing lists of e-addresses. The most exhaustive compilation of e-mail addresses of astronomers was compiled by Benn & Martin (*e.g.* 1990).

On the amateur/grand public front, a couple of compilations deserve to be mentioned such as those made available online by *Sky Publishing Corp.* [15] and *Loch Ness Productions* [16].

[15] `http://skyandtelescope.com/resources/organizations/`
[16] `http://www.lochness.com/lpc/lpc.html`

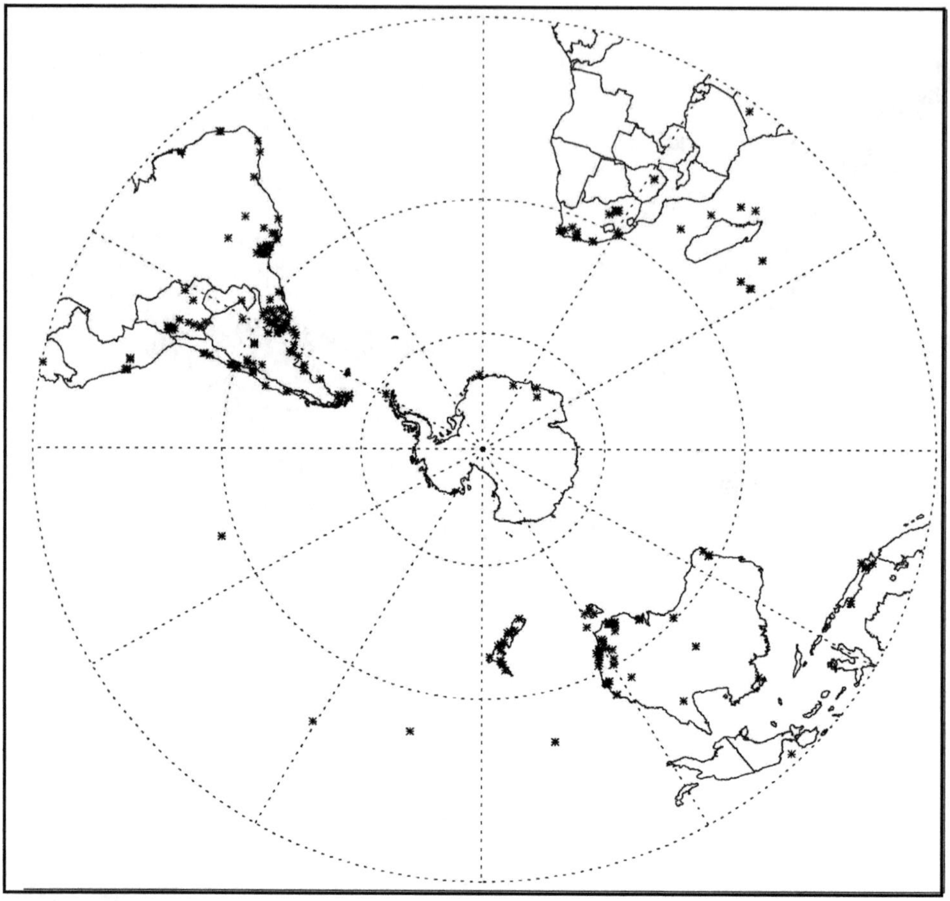

Figure 6. Planet 'Astronomy': Southern countries (Heck 2000a).

5.2. ASTROWEB

Among the purely web-oriented resources, *AstroWeb* deserves a special mention.

In the early nineties, people here and there were also collecting URLs (of institutions, of people, of projects, ...) for their own usage, but also for the benefit of their colleagues. Several of these pionneers[17] joined efforts and set up the *AstroWeb* Consortium (Jackson *et al.* 1994).

[17]Originally from Mount Stromlo and Siding Spring Observatories (MSSSO), National Radio Astronomy Observatory (NRAO), Space Telescope – European Coordinating Facility (ST-ECF), European Southern Observatory (ESO), Space Telescope Science Institute (STScI), Strasbourg astronomical Data Center (CDS), Strasbourg Astronomical Observatory.

Nowadays, if some of the original institutions are *de facto* not contributing anymore to the resource, others are offering mirrors such as the Cambridge Institute of Astronomy and ESA's Villafranca del Castillo Station (Vilspa). Each of the institutions supports a version of the *AstroWeb* service[18]. The separate versions have different styles and contents, but all are computed from the same master database, which is coded in an agreed interchange format.

Each resource record is categorized and many resources have a paragraph describing the resource and containing links to other records. Different presentations of the master listing are available, sorted by category, Internet domain, protocol and name, and name. There is also a searchable version of the merged resource listing, using a WAIS index. The database has been designed to facilitate distributed maintenance.

Specific on-line facilities have been made available for spontaneous contributions. For instance, there are forms by which new resources can be added (*AstroWeb Resource Entry Form*) and existing resources can be edited (*AstroWeb Database Correction Report Form*).

At the time of writing, *AstroWeb* offers slightly more than 3000 distinct records on the CDS server.

5.3. ASTROWEB VERSUS THE STARPAGES

Although people are not comparing similar things, a question sometimes put to us is whether the *StarPages* are related to *AstroWeb* and/or which resource is best. As explained above, the master files for the *StarPages* largely predate and offer much more, and more diversified, information (admittedly in a less visible way) than *AstroWeb*. The two resources are also structured and operated quite differently.

AstroWeb is basically a list of commented URLs: about 3,000, to be compared with the more than 12,000 URLs offered by the *StarPages* through the resources *StarHeads* (for professional astronomers and related scientists) and *StarWorlds* (for astronomy-related organizations, institutions, associations, companies, and so on). But this latter resource is more than just a list of URLs. As shown in Sect. 3.2, *StarWorlds* is a directory with all practical data systematically compiled, authentified and verified. It also includes quite a number of organizations not yet on the web.

[18]Here are the various URLs/mirrors:
`http://cdsweb.u-strasbg.fr/astroweb/consortium.html`
`http://www.mso.anu.edu.au/astronomy/astroweb/astronomy.html`
`http://fits.cv.nrao.edu/www/astronomy.html`
`http://www.stsci.edu/net-resources.html`
`http://www.vilspa.esa.es/astroweb/astronomy.html`
`http://www.ast.cam.ac.uk/astroweb/yp_astronomy.html`

The fact that, in *AstroWeb*, the material is presented in long lists gives not only a feeling of plenty, but allows a direct visible search through it. In the *StarPages*, the full master files are not made accessible as a protective measure for the individuals listed and against excessive download of material, but also to allow usage of more flexible search engines.

In *StarWorlds*, the material retrieved is presented in an homogeneous way, with all practical information available on the various organizations matching the search (see Sect. 3.2). This is the result of daily maintenance, updating campaigns, and inclusion of validated information from signed and documented questionnaires.

In *AstroWeb*, and because of the basically spontaneous on-line submission of URLs by third parties, the level is definitely heterogeneous, not exhaustive, and sometimes questionable. Since they are submitted by the entries themselves, the corresponding presentations are basically self-promoting and thus, in some instances, lack the objectivity desired in a scientific resource.

AstroWeb contains however some URLs (essentially of specific experiments or projects) that cannot naturally fit within the *StarPages*. On the other hand, *StarWorlds* includes full data on a significant amount of entries still without Internet presence and/or web sites. A number of URLs from the *StarPages* were downloaded into *AstroWeb* to help it taking off. More details can be found online[19].

6. The Future

How can be seen the future of yellow-page services in astronomy? Nowadays excellent search engines such as `google`[20] or `AltaVista`[21] allow to retrieve quickly and efficiently all kinds of information pieces – as probably everyone has already experienced. A list of URLs are returned with excerpts of the corresponding pages where the queried words or expressions are appearing.

There is however matter for insatisfaction: the search engines' databases are not updated daily and the information retrieved is not always the latest one; or the links retrieved point to dead pages or to pages the contents of which have been modified since the last scan by the search engines' robots (and have become unrelated to the query). In our view however – from intense continual practice of the web – such shortcomings are also frequently resulting from the carelessness of webmasters and page maintainers: the contents of the pages are not adequately updated; forward links are not inserted from old pages towards new ones, and so on.

[19] `http://vizier.u-strasbg.fr/~heck/awsp.htm`
[20] `http://www.google.com/` + national variants.
[21] `http://www.altavista.com/` + national variants.

The search engine performance depends also

- on the way requests have been formulated, as well as
- on the way web pages have been formulated and structured.

Still, it is amazing how much detail can be retrieved through well-targeted requests. Certainly not everything is available on the web and might never be, especially for historical documents, but most dynamical organizations (like many astronomy-related organizations, isn't it?) have now a web presence that can be detected easily. A decade after the web celebrated birth, its pages have reached a fairly good level of maturity.

Since we astronomers (and our librarians/documentation specialists) have generally become well-trained information providers and consumers, one could wonder about the need to maintain specific lists of URLs (commented or not) since the corresponding pages are easily retrieved through search engines. Such lists of URLs have multiplied since the web's early days – often copying each other. But nowadays they can only stay steps behind the web evolution and the search engines' capabilities.

The situation is different with value-added resources where the critical analysis of the information provided, the authentication of the sources, the homogeneous and systematic presentation, and so on, are part of an irreplaceable maintenance. As Gell-Mann (1997) expressed it: "We hear that, in this dawn of the so-called information age, a great deal of talk about the explosion of information and new methods for its dissemination. It is important to realize, however, that most of what is disseminated is misinformation, badly organized information or irrelevant information." Hence the need to provide good-quality information, relevant to the matter under consideration, information exactly on target and updated whenever needed.

We thus believe that resources such as the *StarPages* have their slot in the future since those properties are exactly what they are aiming at. The corresponding investment in time, updating expenses, and so on, are worthwhile and justified. Gell-Mann added though: "How can we establish a reward system such that many competing but skillful processors of information, acting as intermediaries, will arise to interpret for us this mass of unorganized, partially false material?"

This is where the whole sociology is still unsatisfactory since, in terms of yellow-page services, *ad hoc* rewarding schemes have still to be formulated and structured (see also the following section).

7. A Couple of Final Comments

7.1. A WAY PAVED BY FRUSTRATION?

What should be said to a young scientist who would like to initiate such a compilation activity? The first thing is that this should be done with the highest quality possible or not at all. The second point would be that this should be a secondary activity with however substantial amount of time available for careful maintenance. The third condition would be that the job be done by the scientist him/herself and not by being delegated to unexperienced clerks or technicians.

All this must sound rather obvious.

Less obvious is probably the fact that this young scientist should not expect much of recognition or credits, in any, nor a boost for the career. Quite the opposite, he/she should have strong shoulders as criticisms will abound. Users generally see the tiny piece of information incorrect or missing, get upset about it, shoot an e-mail without thinking it twice, forgetting about the human at the other side that will get in full face the anger expressed – strenghtened by the e-mail effect. There is usually no consideration for the vast amount of correct data and the freshly updated material that remain unnoticed.

Fights can take place between the organizations listed, some of them being unhappy by and requesting modification of the information published on the other ones that they find too flattering. Hence the need for the compiler, on one hand, to use sometimes referees and, on the other hand, to protect his/her back by publishing only authenticated information from signed questionnaires or updating forms[22].

Individuals listed in databases behave sometimes as *prime donne* and surrealistic talks have then to be expected, when not hot debates involving occasionally direct personal insults. Take a strong dosis of patience, diplomacy and psychology, but be ready to be tough and to put things back into their right place when they go too far.

Hackers and undelicate people might play nasty tricks. The compiler might see thousands of hours of careful work being sucked away in a few seconds (see also Sect. 7.2 on copyright protection).

The young compiler might also see magazine people who are not doing properly their homework praising impressive lists built up by automatic procedures and of dubious quality while his/her own meticulous work will be overlooked.

It is indeed much easier to carry out one-time job such as devising automatic procedures and to forget later on about the real nature and the quality of the material dealt with. It is quite a different job to maintain daily that material as exactly and as precisely as possible. To compile a directory of real value is quite a different venture to just reproducing and distributing, with comments of greater or lesser interest, data collected indiscriminately and/or automatically from all available source. If professional file construc-

[22]Something not necessary for mere lists of URLs since the contents of the pages pointed at are not the responsibility of the compiler.

tion techniques are necessary, they cannot spare the extensive background, unrewarding and very careful work which is indispensable for the compilation of a valuable directory. The definition of a very well profiled and adapted questionnaire, the homogenization of the data collected and the maximum reduction of the respondents' biases are all points that must be satisfied.

Moreover, it is imperative to take into account national differences: standards, conventions, habits, publication and financing channels usually vary from country to country. Finally, it is scarcely necessary to add that everything relating to professional astronomy can only be dealt with by professional astronomers, for evident reasons of competence and better knowledge of the realities of their corporation.

The data provided by the respondents are not always reliable, especially in the case of amateur organizations. Some groups have indeed a tendency to exaggerate their importance. *Ghost* and dead groups have also to be detected. Organizations giving no sign of life for some time (for instance by not answering questionnaires and update requests) have to be rejected as well as those not presenting all guarantees of seriousness. Refer also to the discussion on automation and quality in Heck (2000b).

All in all, considering the difficulties (when not easy criticisms instead of encouragements) to produce a good-quality job in the field, with virtually non-existing rewarding schemes, it should not be too surpizing few people dedicate themselves to this type of work or others who started it "throw in the sponge" after a while.

7.2. COPYRIGHT PROTECTION OR ANARCHY?

A series of hacking incidents with one of our databases (*StarBits*) showed how fragile was all that compilation work done over so many years and how little efficient was the support received in spite of the good will of parties involved.

It is obvious that hackers and pirats will always be ahead of electronic security setters, just like in crime (and this is crime, after all). Legal protections ("copyright") are nice in theory, but what can they prevent in practice? Huge databases with real or potential commercial interests can be downloaded by hackers in a matter of seconds. This is perhaps the most dramatic fragility of the electronic material.

Be also aware that, in some countries, it is enough to alter or to change "significantly" a file, a list or a directory to escape prosecution for infringement of copyright restrictions. Everything is of course in the interpretation a judge will make of the word "significantly".

The protection of material on paper has a longest history and commercial publishers have well-experienced lawyers to deal with the "copyright" of their products. Still, a directory, including many addresses, attracts unavoidably commercial attention. Our own experience is that most people are respectful of copy restrictions and would get in touch with the right owners whenever necessary. Juicy law suits, especially in the US, have got people think twice before making a wrong step. In Europe, the culture of copyright is less developed (and the suits are still less juicy). We could however settle appropriately, before reaching the courts, various incidents experienced with undelicate commercial companies.

How can one "catch" wrongdoers? Simply by inserting detectors in the files. All the master files for the *Star*s Family* resources and the *StarPages* in particular are seeded with detectors. A standard user would not notice them, but they very efficiently allow to spot quickly illicit copying of the files or wrong usage of the information they provide. One has of course to be imaginative when devising such detectors. Recipes will not be given here as the *StarPages* are operational resources. Be aware that, in some countries, detectors in files must be declared legally. Of course, they are not made public. This serves only to prove the authenticity of detectors before the courts if needed.

We also monitor carefully the uage of the *StarPages* in real time and, thanks to François Ochsenbein from CDS, automatic processes have been designed to kick out and to deny access to users whose queries are excessive or not "normal" (such as systematic scanning). Still, smart people can manage to go round always stricter securities.

Finally do not believe misbehaviors are the privelege of hackers or of commercial companies only. Scientific institutions can also "easily forget" to give appropriate credits and acknowledgements.

References

1. Albrecht, M.A. & Heck, A. 1993, StarGates and StarWords – Online Yellow-Page Directories for Astronomy, *ESO Messenger* **73**, 39-40.
2. Albrecht, M.A. & Heck, A. 1994a, StarWords – A Database of Abbreviations, Acronyms and Symbols in Astronomy, Space Sciences and Related Fields (Announcement of a Database), *Astron. Astrophys. Suppl.* **103**, 471.
3. Albrecht, M.A. & Heck, A. 1994b, StarGates – A Database of Astronomy, Space Sciences and Related Organizations of the World (Announcement of a Database), *Astron. Astrophys. Suppl.* **103**, 473-474
4. Benn, Ch. & Martin, R. 1990, Electronic Mail Guide and Directory, Roy. Greenwich Obs., Herstmonceux (electronic files).
5. Berners-Lee, T.J., Cailliau, R., Groff, J.F. & Pollerman, B. 1992, World-Wide Web: The Information Universe, in *Electronic Networking: Research, Applications and Policy*, 52-58.
6. Berners-Lee, T.J., Cailliau, R., Luotonen, A., Nielsen, H.F. & Secret A. 1994, The World-Wide Web, *Comm. ACM* **37-8**, 76-82.

7. Eichhorn, G. *et al.* 2003, The Development of the Astronomy Digital Library, this volume.
8. Gell-Mann, M. 1997, Lecture to ACM97, San Jose, CA.
9. Hardin, J. 1993, Human Collaborations Technologies for the Internet – NCSA Mosaic and NCSA Collage, Communication to *Astronomical Data Analysis Software and Systems III*, Victoria, BC, 13-15 Oct. 1993.
10. Heck, A. 1992, Introducing ... the Star*s Family, *CDS Inform. Bull.* **40**, 139-141.
11. Heck, A. 1993a, StarGuides – A Directory of Astronomy, Space Sciences and Related Organizations of the World (Announcement of a Catalogue), *Astron. Astrophys. Suppl.* **102**, 85-86.
12. Heck, A. 1993b, StarBriefs – A Dictionary of Abbreviations, Acronyms, and Symbols in Astronomy, Space Sciences, and Related Fields (Announcemnt of a Catalogue), *Astron. Astrophys. Suppl.* **102**, 87.
13. Heck, A. 1995, StarHeads (Announcement of a Database), *Astron. Astrophys. Suppl.* **109**, 265.
14. Heck, A. 2000a, Characteristics of Astronomy-Related Organizations, *Astrophys. Sp. Sc.* **274**, 733-783.
15. Heck, A. 2000b, From Data Files to Information Hubs: Beyond Technologies and Methodologies, in *Information Handling in Astronomy*, Kluwer Acad. Publ., Dordrecht, 223-242.
16. Heck, A., Ciarlo, A. & Stokke, H. 1992, StarWays – A Database of Astronomy, Space Sciences and Related Organizations of the World (Announcement of a Database), *Astron. Astrophys. Suppl.* **96**, 565-566.
17. Heck, A., Egret, D. & Ochsenbein, F. 1994, StarWorlds – StarBits (announcement of two databases), *Astron. Astrophys. Suppl.* **108**, 447-448.
18. Heck, A. & Manfroid, J. 1977, Atlas Photographique Astronomique, Éd. Desoer, Liège, 224 p.
19. Jackson, R., Wells, D., Adorf, H.M., Egret, D., Heck, A., Koekemoer, A. & Murtagh, F. 1994, AstroWeb – A database of links to astronomy resources (announcement of a database), *Astron. Astrophys. Suppl.* **108**, 235-236.
20. Klare, G. 1996, Porträtgalerie, Astronomische Gesellschaft, Hamburg, 118 pp.
21. Lukac, M.R. & Miller, R.J. 2000, List of Active Professional Observatories, *USNO Circ.* **178**.
22. Pecker, J.Cl. 1979, Report IAU Comm. 5 on Documentation, in *Reports on Astronomy – Vol. XVIIA – Part 1*, Ed. E. Müller, D. Reidel Publ. Co., Dordrecht, 7-11.
23. Rigaux, F. 1959, Les Observatoires Astronomiques et les Astronomes, Obs. Roy. Belgique, Bruxelles, 460 p.
24. Rigaux, F. 1961, Les Observatoires Astronomiques et les Astronomes – Supplément, Obs. Roy. Belgique, Bruxelles, 42 p.
25. Stroobant, P., Delvosal, J., Philippot, H., Delporte, E. & Merlin, E. 1907, Les Observatoires Astronomiques et les Astronomes, Hayez, Bruxelles, 318 p.
26. Stroobant, P., Delvosal, J., Delporte, E., Moreau, F. & Vanderlinden, H.L. 1931, Les Observatoires Astronomiques et les Astronomes, Casterman, Tournai, 316 p.
27. Stroobant, P., Delvosal, J., Delporte, E., Moreau, F. & Vanderlinden, H.L. 1936, Les Observatoires Astronomiques et les Astronomes – Supplément, Duculot, Gembloux, 106 p.

PRE-COLLEGE ASTRONOMY EDUCATION IN THE UNITED STATES IN THE TWENTIETH CENTURY

J.E. BISHOP

Westlake Schools Planetarium
24525 Hilliard Road
Westlake, OH 44145, USA
jeanbishop@aol.com

> *"It has been said that we may judge of the degree of civilization of a nation by the provision which the people of the nation have made for the study of astronomy."*
>
> W.W. Campbell, Lick Observatory, 1920.

Abstract. The nature of pre-college astronomy education in the United States can be divided into several periods: 1900 to about 1955, 1955 to about 1980, and about 1980 to 2000. Until the Space Age, astronomy in elementary and secondary schools was minimal, a situation influenced in great part of the work of the National Education Association Committee of Ten in 1892. With the launch of the Russian Sputnik in November 1957, a rapid response of concern and action took place to improve science and math education, including astronomy. Efforts by small planetariums and the National Aeronautics and Space Administration (NASA) played large roles in re-introducing astronomy back into schools in the 1960s and 1970s. During the last decades, educational-research-based astronomy programs and a nationwide effort to improve astronomy and other science education were important at all pre-college levels. Although the basic astronomical literacy of students leaving secondary school at the close of the century needed improvement, awareness of astronomical discoveries had increased since the opening of the Space Age.

A. Heck (ed.), Information Handling in Astronomy – Historical Vistas, 207-231.

1. Introduction

In the nineteenth century prior to 1895, astronomy was a regular part of the education of most students who finished a United States (US) high-school education. The typical academy and early high-school curriculum included at least one course with the elements of astronomy. Teacher preparation schools, known as "normal schools", included astronomy as part of their four-year courses and as sections of physical geography and natural philosophy within their two- and three-year courses (Commissioner of Education 1898-99). A number of colleges and universities required high-school astronomy for admission, and astronomy was a basic upper-level course in a liberal college education.

Astronomy educational interest was no doubt advanced as a wave of observatory construction was extended into high schools and teacher-preparation colleges (Warner 1976). The tradition of astronomy in institutions of higher learning from Colonial times was a legacy from England where it was one on the medieval quadrivium in the Liberal Arts course (Gross 1962). Scholars communicated their astronomical knowledge in the colleges and normal schools, and a strong tradition of teaching and learning astronomy was in effect. A study of science in grades 7 and 8 in the nineteenth century reveals that astronomy was emphasized (Rumble 1941). Numerous written anecdotes show that elementary teachers and parents were aware of the sky and could give informal education to young children.

But the opening of the Twentieth Century saw a rapid decline in pre-college astronomy education, which was to continue until the dawn of the Space Age (Bishop 1977, 1979, 1980b, 1981 & 1990). This decline began at the very time that the United States was achieving international status as George Ellery Hale and others made strides in the young field of astrophysics and while many new observatories were turning out huge amounts of positional and celestial mechanics data.

I have identified a major cause of the decline in astronomy education at the beginning of the century as work of a very influential body in 1892, The Committee of Ten (US Bureau of Education 1893; NEA 1893; Bishop 1977, 1979, 1980a, 1981 & 1990). Because courses, topics and methods varied immensely from school to school, colleges and universities were appalled. Universities needed to give alternatives; one determination is that there were 2,760 ways to enter Harvard University (Sizer 1964)! To understand what happened in astronomy education during the first half of the twentieth century, it is important to understand the role of the Committee of Ten.

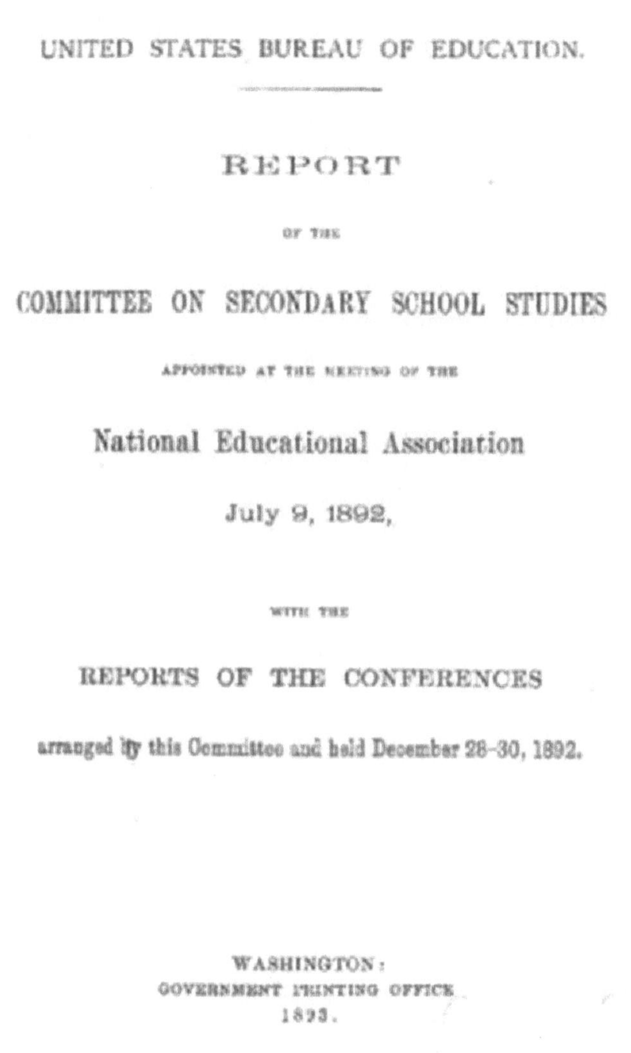
UNITED STATES BUREAU OF EDUCATION.

REPORT

OF THE

COMMITTEE ON SECONDARY SCHOOL STUDIES

APPOINTED AT THE MEETING OF THE

National Educational Association

July 9, 1892,

WITH THE

REPORTS OF THE CONFERENCES

arranged by this Committee and held December 28-30, 1892.

WASHINGTON:
GOVERNMENT PRINTING OFFICE
1893.

Figure 1. The title page of the report of the Committee of Ten of 1892 issued by the National Educational Association. This Report had a strong effect on high school science, influencing near-disappearance of astronomy in US pre-college schools for over 50 years.

2. The US NEA Committee of Ten of 1892

By 1888 Harvard University President and past chemistry professor Charles W. Eliot, with other college administrators, became very concerned with inconsistent preparation for college. In the limelight for his promotion for reform, in 1892 Eliot was appointed by the US National Education Association (NEA) to resolve the diversity problem. Eliot assembled and became chair of the Committee on Secondary School Studies. The popular name for this group then and since that time is the Committee of Ten. The Committee of Ten of 1892 appointed nine additional subject subcommittees, which each held conferences and made recommendations for the high-school curriculum. One subcommittee/conference was designated "Physics, Astronomy, and Chemistry". This conference was chaired by Ira Remsen, chemist and writer of a series of chemistry textbooks. The conference included five college professors and five high school teachers, but no astronomers or astronomy teachers. Probably much of the power of the Committee of Ten was due to the situation "the members they chose for the conference mirrored the Committee's own biases" (Sizer 1964, p. 118).

The members of the Physics, Astronomy, and Chemistry Conference switched the importance of astronomy in its title by becoming "Physics, Chemistry, and Astronomy". They significantly declared that astronomy should not continue to be required for college admission, but physics and chemistry should be required. They endorsed 50 experiments for high school physics and 100 experiments for high-school chemistry. They said that in the future astronomy was to be merely a 12-week elective and did not make recommendations for astronomy course topics. Neither did this subcommittee mention astronomy topics that could be suitable for elementary school, while at the same time the Natural History subcommittee mapped out detailed recommendations for biology in elementary grades that were followed for the next 50 years. Any place for astronomy in elementary education that might be inferred from the work of the total Committee of Ten is found in the recommendations of the Geography Conference, which included detailed subject matter dealing with Earth motions. The Report of the Physics, Chemistry, and Astronomy Conference was only four-and-one-half pages in length, the briefest of all the reports (US Bureau of Education 1893).

Marche (2002), following suggestions by Bishop (1977), emphasizes that educational and social philosophy were changing at the beginning of the twentieth century, and this had an important effect on astronomy education. The theory of mental discipline, the discipline of a subject in training a student's mind, had influenced education in the nineteenth century. At the end of the nineteenth century the mental discipline ideas was challenged with a belief that it was more important to train students for general life experiences. The members of the Committee of Ten seemingly were influenced by these conflicting philosophies, and the country that accepted their recommendations also surely was influenced them.

The impact of the Report of the Committee of Ten on US astronomy education was great, as can be observed by these events:

1. All subject recommendations of the nine conferences were discussed

and debated at educational meetings, large and small, and in hundreds of educational journal articles and popular magazine reviews, following the printing of 30,000 copies of the Report printed by the Secretary of the Interior and 15,000 more copies by the American Book Company and acceptance of the Report at the July 1894 meting of the National Education Association. American popular periodicals, *Harper's Weekly* and *The Atlantic*, carried debates (Sizer 1964).

2. Many content and methods recommendations for the different high-school subjects and elementary levels were soon adopted by both elementary and secondary schools.
3. The laboratory methods promoted by the Committee of Ten were seen in books and articles in science education journals beginning very soon after the Report's publication, including several astronomy laboratory manuals. One manual refers to the Committee of Ten's recommendations in the preface.
4. The elective subject system emphasized by the 1892 Committee thrived in the twentieth century. This was true in spite of the treatment of science electives of an influential 1895 National Education Association Committee appointed to refine the Committee of Ten's recommendations. This Committee on College-Entrance Requirements essentially rubber-stamped the Committee of Ten's core-subject science recommendations by default, and it omitted the electives recommended by the Committee of Ten in their report (Marche 2002). The principle of science electives of the 1892 Committee of Ten survived without this committee's endorsement, although subjects suggested as electives (geology and physiology) experienced declining enrollments in and after this period.

   Astronomy, recommended as an elective by the Physics, Chemistry, and Astronomy Conference, without a supporting list of activities like those given for physics and chemistry, could not compete with physics and chemistry for school system funds. Astronomer Asaph Hall, at the end of the nineteenth and opening of the twentieth centuries, said that a number of high-school astronomy teachers told him that they were not able to obtain money for equipment because their subject could not be listed by students for college admittance (Hall 1900).
5. The credit system for admittance to college recommended by the Committee of Ten was adopted and remains in effect even today, following large changes in science curriculum in the second half of the twentieth century.
6. Junior high schools (grades 7 and 8) flourished after the Committee of Ten recommended that secondary studies be started two years earlier.
7. US Office of Education Reports show for the years 1895-1910 a 90%

decrease of college-bound students in astronomy enrollment but only 30% decreases in physics and chemistry.

Astronomy, because it was no longer required for college while so many other courses were required for college, was squeezed out of high-school and teacher college curriculums. Because astronomy had been eliminated as a college-admittance requirement, high-school students took chemistry and physics rather than astronomy, although a decrease in number of students enrolling in all physical sciences occurred, attributable to an atmosphere of greater societal pressure for education for life instead of mental discipline. Astronomy disappeared from list of offerings at teacher preparation institutions. This decline phenomenon can be seen in the enrollment figures of the US Office of Education Reports (1894-95 to 1930) and Bulletins, and it was noted by teachers and others (Hutson 1927).

A review of US Office of Education Reports reveal that in 1895, 5% of high-school students took a full course in astronomy. By 1905, only 1.2% took such a course and by 1930 only 0.06% of all high-school students studied astronomy. Speaking with a perspective of 70 years, Theodore Sizer stated, "The Report of the Committee of Ten has had a profound influence in American education. For fifteen years after its publication it served as gospel for the curriculum writers of the burgeoning high schools (Sizer 1964, p. xi).

Educational programs initiated in Puerto Rico in the early 1900s mimicked the science education trends in the United States (Lindsay 1905). At an AAS conference in the 1950s, an astronomer from Puerto Rico mentioned in conversation that astronomy was not offered in Puerto Rican high schools because of the "negative work of the Committee of Ten of 1892" (Emmons 1975).

Most US astronomers did not attempt to change this trend. US Naval Observatory Director astronomer Simon Newcomb was chair of the Committee of Ten's Mathematics Conference, but he did not publish his comments about any decisions of the Physics, Chemistry, and Astronomy Conference. His former student Edward Holden, Lick Observatory Director, prepared a report on both pre-college and university astronomy education, published in 1899.

Of like mind with the Committee of Ten, Holden found no fault with decisions of the Committee of Ten and wrote that high school students planning to become astronomers should get a broad educational foundation and leave astronomy for college (Marche 2002). The American Astronomical Society (AAS) was formed in 1899 with Simon Newcomb as its first President (Sagan 1974), but when its first education group was formed in 1911, their focus was on the nature of astronomy courses at colleges and universities (Fraknoi & Wentzel 1997).

Figure 2. "Jane and Ned learning the names of some of the constellations. Knowing a few stars and constellations has kept many people from getting lost at night". Illustration and caption from Pathways in Science: Learning About our World (Book 5, p. 146), published by Ginn and Co. in Boston, 1932. This is one of the popular graded elementary textbooks of the 1930s. This particular book had a better astronomy study section than most science readers of the time; few had more than brief references to astronomy. Notice that the picture and caption reflect the priorities of a life-adjustment philosophy prevalent at this time.

Astronomers Mary E. Byrd, William Tyler Olcott, and Asaph Hall tried to convince pre-college educators that more astronomy was needed, Olcott with his 1907 guidebook A Field Book of the Stars (Marche 2002; Hall 1900). But their work did not reverse the decline in US pre-college astronomy education that was to persist for some time. One may wonder what effect the inspired and charismatic contemporary books of French astronomer Camille Flammarion would have had if they had been read at this time by influential US educators (Bishop 1981).

3. 1900 to 1955

The cycle of astronomy education that had made possible the knowledge of rudimentary astronomy by most people, including informal teaching by parents as well as by teachers in schools, vanished following the events surrounding acceptance of the 1892 Committee of Ten's recommendations. For about 50 years the average educated US citizen largely was illiterate in astronomy.

In early decades of the new century, nature study was emphasized in elementary grades. Specific topics and methods recommended by the Committee of Ten's Natural History subcommittee were followed in many schools. Occasional rapid surveys of the known universe, not reflecting current astronomical understanding, were found in the author's survey of 15 elementary readers used at this time. The explanations of every-day astronomical phenomena, known to so many in earlier generations – why the stars rise and set, the observed changing noontime altitude of the Sun an Earth revolution, the motions of the moon and planets – were not included.

It was apparent to educators in the first half of the twentieth century that science strongly affected human lives, so elementary school science emphasized science in daily living and technology related to science. A set of precepts based on writings by educator John Dewey burst forth in various documents that were affirmed in 1918 with the title of the Cardinal Principles and reaffirmed in 1937, 1944, and 1952 with other names. These ideas guided all pre-college education in this period: health, command of fundamental processes, worthy home membership, vocation, civic education, worthy use of leisure, and ethical character. By 1950 problem-solving related to daily activities, integrated science topics and general science were main features in kindergarten through ninth grade (Bishop 1981).

My survey of readers and textbooks of this period show that astronomy was absent or negligible in these subject organizations, while biology and chemistry topics were emphasized. Some books promoted constellations for their utility, *i.e.* knowledge that aids orientation or further observations. However, astronomy in elementary school books of the first half of the century was superficial, often misleading, frequently outdated, and sometimes false.

In 1915 an astronomy review noted , "To a utilitarian age astronomy seems a somewhat worn out, useless science: it has been relegated to a few pages in schoolbooks on geography, pages that are doubtless often omitted when they are studied" (Jacoby 1915, p. 258). Elementary geography books usually show a limited introduction to Earth rotation and revolution, which often by omissions seem a fertile source of misconceptions. In 1920, at a dedication address of Cleveland, Ohio's Warner and Swasey observatory, W.W. Campbell lamented, "the universities, the colleges, and the technical schools of our country, and of other counties, are graduating every year hundreds of young men, ready to start upon the more serious phases of their lives, who can tell us all about the lights in our houses, but not one word about the lights in our sky (Campbell 1920). By the late 1920s, only 39 US high schools offered a course in astronomy (National Society for the Study of Education 1932).

Popular education provided information for a small percentage of the pre-school population in the 1930s. An amateur astronomy movement ap-

Figure 3. A color illustration in the World Book Encyclopedia (Volume 1), published by the Quarrie Corporation of Chicago, demonstrating distance to planets by noting the time to travel to each in a plane traveling at 100 miles per hour. This is from the 1937 edition. Notice that Pluto, discovered by Clyde Tombaugh in 1930, is missing from the illustration. Popular material, from which students as well as adults learned their astronomy during this time, frequently kept old plates, and astronomical discoveries had to be gleaned from news releases and not read in books for some years.

peared in this decade and *Popular Astronomy* magazine was an important source of information for students hungry for astronomy information (Warner 1976). High-school science clubs, a Boy Scout Merit Badge in Astronomy, the appearance of some popular books covering observational astronomy and current astronomy research topics were signs of this movement (Bishop 1981).

One book that conveyed information and excitement of astronomy to adolescents, mentioned to the author by some retired astronomers, was *The Stars for Sam* (Reed 1931). And some scientists, including Hector Macpherson, W.W. Campbell, and E.P. Lewis spoke or wrote publicly about the value of astronomy (Marche 2002; Bishop 1981). The construction of the giant mirror for the Hale telescope on Mount Palomar and its slow passage by rail across the United States from New York to California, also ignited the imaginations of some students for astronomy. Groups of children were released from school during school hours and young children were held in the arms of parents and grandparents to see the precursor of the "giant eye" roll by (Florence 1997).

Large planetariums were opened and began to take an important role in informal astronomy education, influencing young students who attended them within classes and with their parents. The first planetarium in North America, the Adler Planetarium, opened in May 1930, in Chicago, and others soon followed in New York, Philadelphia, Pittsburgh, and Los Angeles.

Excellent astronomy museum areas usually accompanied the planetarium facilities, greatly influenced by excellent astronomy museums found in Munich (The German Museum), London (The Science Museum at South Kensington), and Vienna (The Natural History Museum) (Fisher 1926). In the 1930s and 1940s the large planetarium installations both in the United States and elsewhere were viewed as remarkable educational aids with the specific purpose of disseminating astronomical knowledge Programs were given "live" (not recorded or computerized). Planetariums communicated the astronomy learned by research to people of all ages (Sunal & Sunal 1976).

The majority (60%) of astronomers who received their doctoral degrees between 1945 and 1953 said that their interest in astronomy developed prior to college (Risley 1954). They must have been attracted by informal astronomy education opportunities, since astronomy was missing from almost every school at every level.

It appears that most pre-college students knew almost nothing about astronomy. In 1940 over 300 freshmen at one college were tested on basic astronomy and geology concepts. Over half of the students could not identify that:
a) the Sun is smaller than some other bodies in the universe,
b) meteors are in the atmosphere rather than deep space when we see them,
c) the Sun is a star,
d) the distant stars shine because they are hot like the Sun,
e) the Sun is not in a fixed position in space, and
f) the seasons are not caused by varying distances of the Sun from the Earth.

These students had about the same scholastic aptitude as freshmen in other region colleges and universities, so the authors concluded that the results were representative of the knowledge of many freshmen (Ralya & Ralya 1940).

In 1954 the AAS formed a Teachers' Committee. A column sponsored by the AAS, "Sky and Teacher", conducted by Stanley P. Wyatt, Jr. was published in *Sky and Telescope*, beginning in November 1953 (Wyatt 1953). The Teachers' Committee and column had an objective to assist "all kinds of teachers of all kinds of students" (p. 19). Frequently the column contained resource lists and evaluations helpful to pre-college teachers and informal

Figure 4. In 1954 the book The Exploration of the Moon with 45 imaginative pictures by R.A. Smith and text by A.C. Clark was published by Harper and Brothers in New York. This picture (p. 63), titled "Retrieving Supplies," is one in a sequence that tells a story of a lunar expedition. Books like this helped to excite the generation of students that would work on space efforts in the following decades.

educators. Unfortunately, the Teachers' Committee was short-lived, dissolving when its chair, Carl Bauer, resigned saying "it appeared as though ... members felt that a certain stigma attaches to being asked to do something for the Teachers' Committee" (AAS Council minutes, p. 657, reported by Fraknoi & Wentzel 1997).

The imminent Space Age was not generally anticipated. Educators were surprised when on 29 July 1955, President Eisenhower announced plans for a satellite, and some began musing about the implications for curriculum (Santuosuosso, 1957).

4. The Space Age

The Space Age arrived suddenly and in full force in the fall of 1957. Headlines such as "Ike Views Science Education Tonight" rocked the country on 13 November 1957, as the satellite Sputnik was successfully launched into Earth orbit by the USSR. President Eisenhower said to a receptive population, "I wish that every school board and every Parent-Teacher Association would this week and this year make one single project their order of business. This is to scrutinize your school's curriculum and standards; then decide for yourselves whether they meet the stern demands of the era we are entering". Eisenhower asked people to remember that, when a Russian graduates from high school, he has had, along with other stringent preparation, a year of astronomy (Rosemergy 1978).

When Americans looked at the details of the curriculum in Soviet schools, they were startled and impressed. In a ten-year standardized public school program 42.4% of school time was devoted to math, the sciences and technical practice. Astronomy was studied only 0.3% of time, but it was a stringent course offered once each week in the last year following two years each of physics and chemistry. Physics was studied 5.6% of school time, and chemistry was studied 3.5% of school time (US Office of Education 1957).

Almost immediately it became clear that American schools needed more rigorous science. Abilities of bright students were not being developed, and the country needed their talents to keep up in the "space race". Educational reforms in science education following 1957 came at a dizzy pace. The National Defense Education Act (NDEA) of 1958 and the Elementary and Secondary Education Act of 1965 poured billions of dollars into efforts for improving science education. Astronomy education received an important share of the focus and funding (Rosemergy 1978).

The best known post-Sputnik astronomy curriculum projects are the elementary University of Illinois Astronomy Program, the Earth Science Curriculum Project (ESCP), and the Harvard Project Physics. The elementary program, published in 1969 following eight years of development and testing, was first used with gifted elementary students, and then suggested for grades 6 and 7. The ESCP, with first published materials in 1963, was most frequently offered to students in grades 8 or 9. Harvard Project Physics was intended as a novel, historical approach to high-school physics which would attract students not previously interested in science. Book Two is devoted exclusively to astronomical motions. Recognizing that few teachers had any background in astronomy, elaborate teacher guides were prepared for all three programs. (Bishop 1977 & 1980a).

The science-reform curricula of the 1960s were led by college professors, who felt indignant that educators had allowed scholarship standards to plummet. The scholars inserted a process approach in all the new curricula, judging that knowledge of science required experience with the scientific method and critical thinking. Echoing the beliefs of the curriculum project directors, the Science Curriculum Committee of the National Science Teachers Association (NSTA) of 1959-69 concurred that all science should be studied with the process approach. The Committee determined program standards for kindergarten, with recommendation that aspects of the study

Figure 5. Planetarium Director Dan Snow presents a lesson to a class of elementary students at the Cleveland Museum of Natural History in Cleveland OH in 1952. The instrument is a Spitz A-1. The A-1 model was improved upon by the Spitz projector models A-2 and then the A3P and the A-4, sold in the US by the hundreds in the 1960s and 1970s. (Photograph copy given to Jeanne Bishop by Dan Snow)

of the universe be initiated at every level (Bishop 1977 & 1980a).

In 1969, a separate education conference was sponsored by a new AAS Committee on Education in Astronomy (CEA). CEA Chair George Abell gave a review in his keynote address that included lack of astronomy training by elementary and high school teachers and planetarium staff. He concluded that some of the new science curricula were too difficult for precollege students, and he warned of the need for astronomers to turn attention to and find solutions to the problems (Fraknoi & Wentzel 1997).

By the early 1970s, there were not enough positions for research astronomers. Education seemed an appropriate alternative career. A group led by Bart Bok told the AAS Council in December 1971, that the AAS needed to help both PhD's and MA's obtain positions in smaller colleges, high schools, and planetariums (Fraknoi & Wentzel 1997). A new AAS Task Group on Education in Astronomy (TGEA) formed in 1972, open to anyone who wanted to be active in astronomy education. A large number of education programs were initiated under the leadership of Donat Wentzel and Gerrit Verschuur.

Projects that impacted pre-college education were: Four Brochures for High-School Students on Topics at the Forefront of Astronomical Research, Collections of Astronomy Activities for the Classroom, the Bok Prize for Outstanding High School Projects in Astronomy, A booth at meetings of the National Science Teachers' Association (NSTA), and coordination with other organizations to further astronomy education at all levels (Fraknoi & Wentzel 1997).

One feature of the Four Brochures project was particularly innovative, the opportunity for astronomers and high-school educators to become familiar with aspects of one another's professions. In 1974 a group of about 30 met at the University of Richmond. Together they developed activities for the brochures *Atoms and Astronomy, The Supernova, Chemistry Between the Stars,* and *Extragalactic Astronomy.* The National Science Foundation (NSF) and NASA co-sponsored this project with the AAS (Bishop 1980a).

Other scientific and scholastic groups wrote and distributed a very large number of astronomy education materials. These groups included textbook and educational materials companies, state departments of education, instructional television corporations, amateur science societies, and individual colleges and universities. (Bishop 1980a).

NASA was a very strong contributor to education in the last three decades of the twentieth century. Brochures, books, posters, photographs, films, slides, and teacher activity booklets "gushed forth from the Superintendent of documents and the public education offices of the various centers, arriving in teacher mailboxes with revolving-door frequency" (Bishop 1980a, p. 6). From the late 1960s onward NASA centers sponsored numerous workshops, briefing sessions (before, during, and after launches and data returns), and tours for teachers, planetarium educators and students.

In 1978 scientists at the Lyndon B. Johnson Space Center suggested that lunar rock and soil samples be shared with classroom teachers, and the NASA Lunar Sample Education Project was born. After a required briefing teachers may borrow a disc of six fragments gathered from different lunar sites (Bishop 1980a). Analyzing the wide variety of offerings by NASA for education, Andrew Fraknoi notes that there is an inconsistency in the ways in which NASA materials were distributed . No database seems to exist for the huge supply of materials produced for 30 years (Fraknoi 1998).

When national yearning for knowledge arrived with the Space Age, small planetariums became very numerous, particularly in schools. President Eisenhower's Advisory Council selected the small planetarium as one of the outstanding innovative educational programs of his term of office. The National Defense Education Act (NDEA) of 1958 provided $70 million/year for four years in a program called Title III. Title III money was given on a matching-funds basis to construct hundreds of small planetariums across the country (Bishop 1980a). A study of the *Directory of the World's Planetariums* (1997/98 Edition) reveals that most of the facilities were produced by Spitz, Inc., with the predominant instruments being the A3P and A4 in facilities with fewer than 100 seats.

The state of Pennsylvania was particularly interested in promoting astronomy with new equipment, mandating that any new high school in Pennsylvania must include either a planetarium or an observatory (Bishop 1980b). Large and small museum planetariums joined the educational emphasis, and Joseph Chamberlain, President of Chicago's Adler Planetarium concluded that the planetarium reaches its greatest potential when used for education (Bishop 1980b). The National Assessment of Educational Progress in Science taken between 1969 and 1977 found that more than half of 13-year olds and more than 65% of 17-year-olds in the country had been to one or more educational planetarium programs. Fletcher Watson of the Harvard School of Education found that planetariums reached more students than all astronomy courses and courses in which astronomy was included (Bishop 1980a).

In school districts that had a planetarium, almost always a high-school astronomy course was introduced, bringing back to some high schools what had been missing since the beginning of the century. The course usually was an elective with less status than chemistry or physics, a remaining stigma of the Committee of Ten. A very positive effect of planetariums within school districts was children's increased interest in astronomy. Additionally, elementary teachers learned some basic astronomy and were better able to explain it to children (Bishop 1980a).

5. Student Reasoning and Astronomy

In the 1970s and 1980s, the extensive work of French psychologist Jean Piaget was translated, studied, and applied to astronomy teaching and learning. A primary conclusion from Piaget's work is that children are incapable of thinking in particular ways until certain ages and experience levels. Prior to about age 11, most children cannot think in an abstract way (Piaget & Inhelder 1956; Piaget 1972a&b).

Robert Karplus and others at California's Lawrence Hall of Science developed a method of teaching based on Piaget's work, a Learning Cycle, which was disseminated at workshops at NSTA conferences (Karplus 1972 & 1977; Karplus *et al.* 1977) led to increased interest in research and application in science education., and pertinently, in astronomy education.

Figure 6. This ninth grade student (age 14) puzzles over an activity created by the Earth Science Curriculum Project (ESCP). She must place the stellar spectra strips in some meaningful order. The ESCP text and laboratory materials were widely available for grades 8 and 9 in the 1970s and 1980s. (Photograph by Jeanne Bishop)

An AAS project of the TGEA was the publication of *Effective Astronomy Teaching and Student Reasoning Ability* (Schatz *et al.* 1978), mostly intended for college teachers. Some PhD dissertations in astronomy education were completed based on Piaget/Karplus paradigms (*e.g.* Bishop 1980b; Kelsey 1980), and some articles appeared in teacher magazines promoting the use of Piaget ideas in teaching basic astronomy concepts (Bishop 1992).

6. The Last Decades

In 1988, Northern Illinois University's Public Opinion Laboratory surveyed 2,041 adults. One questions was, "Does the Earth go around the Sun or the Sun around the Earth?" Within this group 72% got the question right, 21% got it wrong, and 7% said they did not know. Of the 72% who got it right, 45% knew how long it took, while the rest did not. With limitations of sampling research, this implies that 94 million Americans do not know that the Earth revolves about the Sun in a year (Fraknoi 1998). Clearly

Figure 7. A high-school student uses a transparent celestial globe and a small model of the solar system to determine the relationship between the perspective from Earth and from a point beyond the solar system. Projective spatial ability, identified by Jean Piaget and important to many astronomy concepts, becomes possible for most students in the final years of high school and college, but not before. (Photograph by Jeanne Bishop)

there was a need for greater astronomical literacy in students graduating from high school. An influential videotape, "A Private Universe", was made in 1987 by Philip Sadler. Viewers see outstanding high-school students and non-astronomy Harvard University graduates not being able to basic concepts such as why there are seasons. Shown to government officials and used heavily in teacher professional development workshop, the videotape introduced cognitive issues to many teachers and scientists (Sadler 1987).

An ambitious project to counteract a felt need for better science literacy was begun in 1989. Project 2061 was widely supported and developed by a large number of government and university scholars and 150 pre-college teachers and administrators, under the initial direction of F. James Rutherford (AAAS 1993). The goal of Project 2061 was to find ways to make all students literate in science, mathematics, and technology by the time they graduated from high school. The guidelines or "benchmarks" are intended as "a powerful tool" (p. VII) for US school districts to structure local curriculums.

The book, *Benchmarks for Science Literacy* (AAAS 1993) lists sets of concepts in each of the sciences that were judged to be appropriate for grades K-2, 3-5, 6-8, and 9-12. In the "Physical Setting", under the heading of "The Universe", a comprehensive set of understandings are expected, which

1) relate to perceptual and thinking abilities at each level,
2) include both basic observational phenomena and some of the most recent research from astronomy,
3) connect astronomy to mathematics and technology, and
4) emphasize some ideas of organization of space bodies and scales of diameters and distances (AAAS 1993).
By the close of the century the "Benchmarks" were referenced in hundreds of articles and were seriously examined by both State Departments of Education planning proficiency tests and by local school districts in each state. By 2000, Project 2061 was promoting "hands-on" or laboratory methods for teaching science, and finding faults in many textbooks (Raloff 2001a&b).

Significant astronomy programs of the last decades all employed laboratory or activity experiences. Project STAR and Project SPICA were both developed at Harvard's Center for Astrophysics (CfA). Project STAR was the only major astronomy curriculum developed for high-school students. Research about student thinking is an essential component of the entire program. STAR was developed between 1985 and 1992 at workshops by about 50 high-school teachers and scholars. Materials were first commercially available in 1992. With a format of many activities, the student is made aware of a personal theory or misconception relating to each activity concept. Although developed as a full course, most teachers have used it as a resource, excerpting particular activities for other courses (Coyle 2002).

Project SPICA, a program for students in grades 2-8, also was developed by teachers and scholars together at workshops, and it also used an activity approach. Overlapping with the development of STAR, led by s, development and training workshops occurred from 1987 to 1994. At the university workshops, 200 teacher "agents" were prepared to disseminate the SPICA materials. Philip Sadler, initial project director, estimates that 18,000 elementary teachers from all states were trained by the agents (Sadler 2002).

Hands-On Universe (HOU), a curriculum project of The Lawrence Hall of Science began testing in the early 1990s and completed a tested high-school curriculum in 1995. HOU has enabled high-school students to request their own observations from professional observatories. Students download telescope images to their classroom computers and use the HOU image processing software to visualize and analyze data. Open-ended astronomical investigations incorporate topics and skills recognized by Project 2061 and new national standards. Students search for undiscovered asteroids and participate in discoveries. Two Pennsylvania high-school students serendipitously found first light for Supernova 1994I while studying spiral galaxies. By 2000, HOU was an international collaboration, with other countries joining the US Project. About 10,000 students, primarily in the US were using HOU in 2000 (Gould 2002; Hands-On Universe 2001).

Figure 8. Cover of the book of recommendations published by the National Academy of Sciences in 1996. (National Academy Press, Washington DC). The National Science Education Standards was, like Benchmarks, a set of guidelines for content reform at the close of the century. Many of the concept standards are similar to those in Benchmarks. Hundreds of teachers, scientists, engineers, college faculty, school administrators, government officials, and parents were involved in the preparation of Standards. Both Benchmarks and the Standards have been used extensively in science curriculum reorganization within local school districts, a process intended by both efforts.

Hands-On Astrophysics (HOA): Variable Stars in Science, Math, and Computer Education was produced by the American Association of Variable Star Observers. HOU was based on the premise that the study of variable stars is an interdisciplinary field connected to different sciences, mathematics, and technology. Students observe variable stars with binoculars, telescopes, and unaided eye, and then analyzed changes in star brightness. AAVSO members of all ages in 45 countries have provided real data to be used by students (Mattei 1997).

The Astronomical Society of the Pacific (ASP) has taken a forceful role in pre-college astronomy education in century's final decades. An activity book, *Universe at Your Fingertips*, was published in 1995. Consisting of many activities appropriate to a variety of grade levels, one editor estimates that over 35,000 educators have used the first edition (Fraknoi 2002).

Project ASTRO has been another important ASP astronomy education initiative. Development began in 1992 with the first cohort of teachers and astronomers trained in its use in 1994. ASTRO links professional and amateur astronomers with 4th–9th grade teachers. After astronomer-teacher partners are trained together at a 2-day workshop, each volunteer astronomer "adopts" a class and visits the class at least four times during the school year. The class does inquiry-based activities, with the astronomer as the advisor. Expanding from its home in the San Francisco, California, region, more than a dozen training locations across the country existed by the end of the century (Fraknoi 2002; ASP 2002).

Further, the ASP has published the *Universe in the Classroom* newsletter since 1984, and also has sponsored almost-yearly Universe in the Classroom teacher workshops. Written for teachers in grades 3-12, in about 1990 the newsletter had an estimated readership of 20,000 pre-college teachers. Each workshop had between 150 and 200 teacher attendees at ASP meeting sites around the country (Fraknoi 2002).

At the close of the twentieth and into the twenty-first century, the following groups also offer special workshops or curriculum materials for pre-college teachers and/or students: Project ARIES of the CfA, the Challenger Center, the Jet Propulsion Lab, the National Optical Astronomy Observatories (NOAO), the National Radio Astronomy Observatory, the Pacific Science Center, the SETI Institute, the Space Science Institute, the Stanford Solar Center, the Star Date program of the University of Texas McDonald Observatory, and the Young Astronauts Program (ASP 2001).

In the final decades, another type of planetarium became important in pre-college astronomy education: the portable facility. A portable has an inflatable dome, and both the packed dome and boxed projector can be moved by one person from school to school. Most of the hundreds in the US are STARLABS, produced by Learning Technologies since 1977 (IPS

Figure 9. Cover of the book Benchmarks for Science Literacy by Project 2061, published in 1993 by Oxford University Press, New York. Project 2061 derived its title from the year that Comet Halley will return. Benchmarks sets forth many detailed facts and concepts which students should know at different grade levels. For grades 9-12, the outcome concepts include the Big Bang theory, the origin of the elements in the Big Bang and in stars, differences in stars, and the technology used to learn astronomy. Solar system concepts are emphasized for grades 6-8.

1997/1998 Edition; STARLAB 2002). The STARLAB, which holds about 25-30 children, is very well suited to discovery activities.

The following observations were provided by an astronomer and three high-school astronomy teachers, and planetarium teachers who taught high-school astronomy and gave school planetarium lessons for both elementary and secondary students, 1970-2000:

1. The percentage of professional astronomers engaged in talks or assistance to pre-college education is 10% or less, and the order of magnitude has remained relatively constant in the last 30 years (Kaler 2002).
2. Since about 1970, high-school graduates seem to be more aware of astronomical discoveries and their significance. However, the actual un-

derstanding of basic astronomy concepts like seasons and lunar phases has not changed greatly (Marshall 2002; Comienski 2002; Andress 2002).

3. The content of high-school astronomy courses between 1970 and 2000 has been strongly influenced by space program discoveries. NASA web sites, articles with beautiful color photographs and diagrams in popular magazines, and brief but frequent television news reports about space events and discoveries have communicated the excitement of astronomy. Student interest is fueled by these sources (Marshall 2002; Comienski 2002; Andress 2002).
4. During 1970-1980 there was a constant visible difference in what high-school students of different academic ability desired from an astronomy course. Top students enjoyed the synthesis of math, physics, and astronomy. Average students were not comfortable with an astronomy course emphasizing mathematical relationships. Average students enjoyed a general, descriptive approach to astronomy, wishing a broad scope of topics from constellations to stellar and galactic evolution without much mathematics (Marshall 2002).
5. Without a planetarium in a school district, an astronomy course probably would not have been offered within the district. Lessons in a school planetarium greatly enhanced the teaching of the high-school courses (Marshall 2002; Comienski 2002; Andress 2002).

7. Conclusions

In conclusion, the Space Age events brought the US out of its Dark Age for pre-college astronomy education, which was begun by the 1892 Committee of Ten. Space mission discoveries have been an important factor in student interest in astronomy. A couple of generations of astronomers, many now retired, became interested in astronomy through informal activities and not in pre-college classrooms of the period 1900-1955. A small but significant percentage of astronomers have been and still are interested in pre-college astronomy education.

Many excellent programs and materials for both elementary and high-school levels have been produced since the 1960s. Student thinking ability, identification of misconceptions, and active, inquiry-based learning are recognized as important components of successful astronomy learning.

The small planetarium has had an important role in returning astronomy as an elective to some high schools, as well as increasing knowledge and interest of elementary students and teachers. Important to every pre-college curriculum, as determined by the esteemed Project 2061 working at the end of the twentieth century, are: basic observational concepts, an awareness of the scale and organization of the universe, and understanding of connections to new technology.

There is still much to be done to develop astronomical literacy in pre-college students at the beginning of the twenty-first century, although projects are in place which give more hope for the future than concerned teachers and astronomers saw at the beginning of the twentieth century.

References

1. American Association for the Advancement of Science (AAAS) 1993, Benchmarks for Science Literacy, Oxford Univ. Press, New York.
2. Andress, R. August 2002, private communication.
3. Astronomical Society of the Pacific (ASP) 2001, National Astronomy Education Projects: A Catalog, `http://www.astrosociety.org/education/resources/naep02.html`.
4. Astronomical Society of the Pacific (ASP) 2002, Project ASTRO: Astronomers and Educators as Partners for Learning, `http://astrosociety.org/education/astro/project_astro.html`.
5. Bishop, J.E. 1977, United States Astronomy Education: Past, Present, and Future, *Science Education* **61**, 295-305.
6. Bishop, J.E. 1979, The Committee of 10, *Sky & Telescope* **March**, 212-214.
7. Bishop, J.E. 1980a, Astronomy Education in the United States: Out from Under a Black Cloud, *Griffith Observer* **44/March**, 2-10.
8. Bishop, J.E. 1980b, The Development and Testing of a Participatory Planetarium Unit Emphasizing projective Astronomy Concepts and Utilizing the Karplus Learning Cycle, Student Model Manipulation, and Student Drawing with Eighth Grade Students, Unpublished PhD Dissertation, Univ. Akron, 514 pp.
9. Bishop, J.E. 1981, US Astronomy Education, 1895-1955: Rejection in a Golden Era of Astronomical Discovery, Unpublished manuscript, Third Prize in the History of Education competition of the National Science Teachers Association (NSTA), 33 pp.
10. Bishop, J.E. 1990, The Committee of Ten, in *The Teaching of Astronomy*, eds. J.M. Pasachoff & J.R. Percy, Cambridge Univ. Press.
11. Bishop, J.E. 1992, Planetarium Methods Based on the Research of Jean Piaget, in *Planetarium: A Challenge for Educators*, United Nations, New York, 21-27.
12. Campbell, W.W. 1920, The Daily Influences of Astronomy, *Science* **December 10**, 543-552.
13. Comienski, J. September 2002, private communication.
14. Coyle, H. September 2002, private communication.
15. Emmons, R. 1975, private communication.
16. Fisher, G.C. 1926, Popular Astronomic Education in Europe, *Science* **63/January 22**, 81-84.
17. Florence, R. 1997, featured in *Observatories: Stonehenge to the Hubble Telescope*, History Television Channel Documentary, Publ. # AAE-42201.
18. Fraknoi, A. 1998, Astronomy Education in the United States, p. 2 of the Introduction, expanded from a paper in *Astronomy Education: Current Developments and Future Coordination*, ed. J.R. Percy, Astron. Soc. Pacific Conf. Ser. **89**, 9. `http://www.astrosociety.org/education/resources/useduc.html`,
19. Fraknoi, A. September 2002, private communication.
20. Fraknoi, A. & Wentzel, D. 1997, Astronomy Education and the American Astronomical Society, in *The American Astronomical Society's First Century*, ed. D.H. DeVorkin, Amer. Astron. Soc., Washington, DC, 194-205.

21. Gould, A. September 2002, private communication.
22. Gross, R.E. 1962, Heritage of American Education, Allyn and Bacon. Boston, MA.
23. Hall, Jr., A. 1900, On the Teaching of Astronomy in the United States, *Science* **12/July 6**, 15-20.
24. Hands-On-Universe 2001, Descriptive brochure and `http://hou.lbl.gov`.
25. Hutson, P. 1927, The Scholarship of Teachers in Secondary Schools, MacMillan, New York.
26. IPS Directory of the World's Planetariums 1997/1998 Edition, ed. S. Laatsch & D.W. Smith, International Planetarium Society.
27. Jacoby, H. 1915, Astronomy, Cyclopedia of Education, 1, ed. P. Monroe, MacMillan, New York, p. 258.
28. Kaler, J. September 2002, private communication.
29. Karplus, R. 1972, Physics for Beginners, *Physics Today* **25**, 36-46.
30. Karplus, R. 1977, Science Teaching and the Development of Reasoning, *J. Res. Sc. Teaching* **14/2**, 169-175.
31. Karplus, R., Lawson, A., Wollman, W., Appel, M., Bernoff, R., Howe, A., Rusch, J. & Sullivan, F. 1977, Science Teaching and the Development of Reasoning, Lawrence Hall of Science, Berkeley, CA.
32. Kelsey, L. 1980, The Performance of College Astronomy Students on Two of Piaget's Projective Infralogical Grouping Tasks and their relationship to Problems Dealing with Phases of the Moon, Unpublished PhD Dissertation, Univ. Iowa, 117 pp.
33. Lindsay, S.M. 1905, The Inauguration of the American School System in Porto Rico, Report of the US Commissioner of Education, p. 293.
34. Marche, II, J.D. 2002, Mental Discipline, Curricular Reform, and the Decline of US Astronomy Education, 1893-1920, *Astronomy Education Review* 1 (1) http://aer.noao.edu?AERArticle.php?issue=1&section=1&article=6
35. Marshall, J. September 1997, private communication.
36. Mattei, J. 1997, Preface to Hands on Astrophysics, American Association of Variable Star Observers, v-vi.
37. National Educational Association (NEA) 1893, Report of the Committee on Secondary School Studies, Reprinted in 1969 in New York by Arno Press and the New York Times.
38. National Society for the Study of Education 1932, The Thirty-first Yearbook, Part I: A Program for Teaching Science. Public School Publishing, Bloomington, IL.
39. Piaget, J. 1972a, Physical World of the Child, *Physics Today* **25**, 23-27.
40. Piaget, J. 1972b, Intellectual Evaluation from Adolescence to Adulthood, *Human Development* **15**, 1-12.
41. Piaget, J. & Inhelder, B. 1956, The Child's Conception of Space, Humanities Press, New York.
42. Raloff, J. 2001a, Errant Texts: Why Some Schools May Not Want to Go By the Book, *Science News* **159/March 17**, 168-170.
43. Raloff, J. 2001b, Where's the Book? Science Education is Redefining Texts, *Science News* **159/March 24**, 186-188.
44. Ralya, Lynn L. & Ralya, Lillian L. 1940, Some Significant Concepts and Beliefs in Astronomy and Geology of Entering College Freshmen and the Relation of These to General Scholastic Aptitude, *School Science and Mathematics* **40**, 727-734.
45. Reed, W.M. 1931, The Stars for Sam, Harcourt Brace, New York.
46. Risley, A.M. 1954, Sky and Teacher: One Hundred and One Astronomers, *Sky and Telescope* **13**, 160.
47. Rosemergy, J. 1978, Roots and Routes, Armand Spitz Banquet Address of the Great Lakes Planetarium Association, October 20.
48. Rumble, H.E. 1941, A Hundred Years in Science Education at the Junior High School Level, *Science Education* **28**, 261-265.
49. Sadler, P. 1987, A Private Universe (videotape).
50. Sadler, P. 2002, private communication.

51. Sagan, C. 1974, The Past and Future of American Astronomy, *Physics Today*, December, 23-31.
52. Santuosuosso, J.J. 1957, What Are the Implications for American Education of the Satellite Proposal in Ike's Speech of July 29, 1955?, *Science Education* **41/1**, 48-54.
53. Schatz, D., Fraknoi, A., Robbins, R. & Smith, C. 1978, Effective Astronomy Teaching and Student Reasoning Ability, Lawrence Hall of Science, Berkeley, CA.
54. Sizer, T.R. 1964, Secondary Schools at the Turn of the Century, Yale Univ. Press, New Haven.
55. Sunal, D.W. & Sunal, C.S. 1976, The Planetarium in the American School Experience, *School Science and Mathematics* **77/3**, 203-213.
56. US Bureau of Education 1893, Report of the Committee on Secondary School Studies Appointed at the Meeting of the National Educational Association, July 9, 1982, with the Reports of the Conferences Arranged by this Committee and held December 28-30, 1892, Government Printing Office, Washington, DC.
57. US Commissioner of Education. Reports of the US Commissioner of Education, Volumes 1894-95 to 1930, Government Printing Office, Washington, DC.
58. US Office of Education 1957, Education in the USSR, Division of International Education, International Educational Relations Branch, Bulletin 1957, no. 14, Government Printing Office, Washington, DC.
59. Warner, D.J. 1976, 200 Years of Amateur Astronomy, Proceedings of Astrocon-76, The Astronomical League's Bicentennial Convention, August 19-22, 86-93.
60. Wyatt, S.P. 1953, Sky and Teacher, *Sky and Telescope*, November, 19.

THE BIRTH AND EVOLUTION OF THE PLANETARIUM

C.C. PETERSEN
Loch Ness Productions
P.O. Box 1159
Groton, MA 02138, USA
`carolyn@lochness.com`

Abstract. The history of the planetarium and star theater is one of technological evolution for planetarium instruments and the facilities in which they are installed. The age of the modern planetarium began in the early 1900s with the creation of the Zeiss opto-mechanical instrument and continues today alongside the development of computer-driven full-dome video systems that bring a different approach to simulating the night sky. A number of companies in the US, Europe, and Japan produce planetarium equipment and programs. Along with improvements in projectors and auxiliary systems, planetarium shows have also changed. Today, many different styles of presentation are utilized in the world's 2,900+ facilities. Planetarium facilities are staffed by professionals whose backgrounds range from education to multi-media production.

1. Early History

The night sky has always been and will remain a source of fascination for humans. The study of the stars is one of the oldest sciences. Even though much of its early history was intertwined with the so-called astrological arts, astronomy thrives today as a science in which observers at any level can excel and make contributions. Depictions of the sky have been with us since earliest times, and humans have come up with various ways to model the look of the sky and the relative positions between celestial objects. There is evidence going back nearly 6,000 years ago of a globe of the sky invented by the philosopher Anaximander. Nearly 2,000 years ago, globes of the Earth featured sky map overlays. At least one such reproduction –

A. Heck (ed.), Information Handling in Astronomy – Historical Vistas, 233-247.

credited to Archimedes in the third century BC – had moving parts that demonstrated the slow movement of planets across the sky. As well, the geometer Eratosthenes created an armillary sphere – a hollow shell featuring key astronomical concepts. Other attempts to model the movement of the stars in a mechanical fashion have been unearthed dating back to the first century BC. Throughout the middle ages inventors continued to perfect astronomical clocks (astrariums) and orreries (which depict the motions of the planets), all in an effort to model the night sky more accurately.

The first facility that would be recognizable as planetarium-type theater was the Gottorp globe constructed in the mid-17th century. It consisted of a 4-meter-wide wooden dome and seated perhaps half a dozen people. The dome had holes drilled into it, permitting light from outside to create the starry effect of the nighttime sky. The last of these constructs was the Atwood globe, built in 1913 for the Museum of the Chicago Academy of Sciences. It was a five-meter dome that could simulate 692 stars. A moveable light bulb represented the Sun, and the ecliptic had apertures drilled along its line to represent the planets.

These systems depended on outside light to reproduce the stars. The idea of using a light inside a dome to create a starfield had its genesis in an instrument invented around 1912 by E. Hindermann in Basel, Switzerland. It was the Orbitoscope, and used springworks to revolve two planets around a central globe depicting the Sun. One of the planets had a small light bulb attached to it. As the mechanism moved, shadows of the other two objects and their movements were projected onto the dome. The Orbitoscope set the stage for nearly a century's worth of development of planetarium technology. It was to have far-reaching repercussions.

2. The Rise of Zeiss and the Modern Age of the Planetarium

E. Hindermann's invention primarily led to the development of a device that would actually project simulations of all the celestial objects onto a dome. In 1919 a design engineer at Germany's Carl Zeiss optical works named Walther Bauersfeld came up with a design for the first planetarium projector that would sit in the middle of a room and recreate the heavens on a white, curved surface. Over the next five years, and with the backing of Oskar Miller, director of Munich's Deutsches Museum, Herr Bauersfeld worked on a mechanism that could replicate the sky onto a dome and also be easily controlled. Accuracy was of paramount importance, and so Bauersfeld and his engineers set out to create what became the grandfather of the modern projection planetarium. For the star plates they used film images of 4,500 stars. Bright bulbs shone through these plates, creating the starfield. The design team figured out a method to connect daily and

Figure 1. Zeiss Mark I. This first modern opto-mechanical planetarium instrument was dubbed "the wonder of Jena" when it was first unveiled to the public on October 21, 1923. It is now on display at the Deutsches Museum in Munich, Germany. (Credit: Carl Zeiss Corporation)

annual motion drives so the planets would stay in proper relative positions against that star field. This first instrument – dubbed the Zeiss Mark I – had its first public showing in October 1923. It featured a 0.5-meter star projector with an incandescent lamp at its center. The surface held 31 separate projectors that held the star plates. Although it didn't allow for changes of latitude, this "proof of concept" was well received and set the stage for the nearly century-long development of planetarium instruments and theaters that followed.

With some changes to the design (including a shift to a two-hemisphere star ball arrangement) the Zeiss planetarium projector became widely recognized as a superior astronomy education instrument. Before the onset of the World War II, Zeiss projectors were installed in nearly two dozen cities in Europe, Japan, and the United States.

The war years were not kind to Carl Zeiss, Inc. In the aftermath, the company was split into two parts – with Carl Zeiss, Inc. headquartered in Oberkochen, West Germany and Carl Zeiss-Jena remaining in Jena. Development of planetarium instruments continued on parallel tracks at both

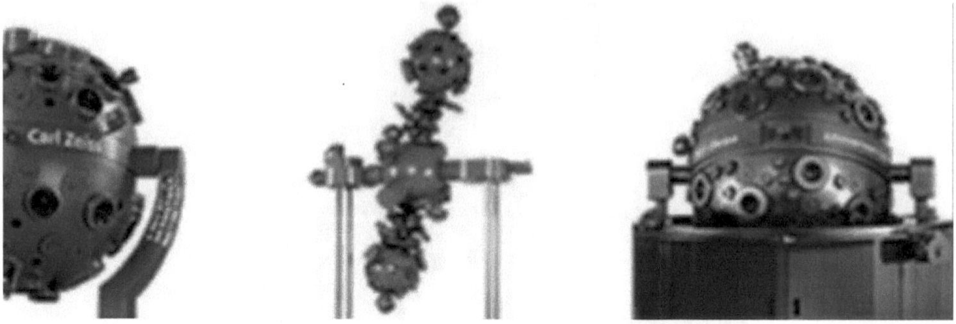

Figure 2. Today's Zeiss instruments. (Credit: Carl Zeiss Corporation)

companies (not surprising, since many of the engineers in both companies had worked together before the war). Sales of Zeiss instruments in the US and much of Western Europe came mainly from the Oberkochen plant, while the Jena operation served Eastern Europe, the Soviet bloc, and other parts of the world. The two parts of the company were reunited in the early 1990s.

3. The Growth of a Technology and a Market

Due to their high quality and precision craftsmanship, Zeiss optics were quite expensive. In 1930 the Zeiss Mark II projector cost in excess of $75,000 – a price only the most well-financed museums could pay. World War II put Zeiss machines out of reach for American facilities for both political and practical reasons. It fell to a lecturer at Philadelphia's Fels Planetarium named Armand N. Spitz to come up with a viable alternative. He founded Spitz Laboratories and in 1947, and after much experimentation in his home workshop, he came up with the Model A, a pinhole-projection planetarium instrument that could be mass-produced and made available for an approximate cost of around $500.00. This was a figure that appealed to a large number of schools, universities, and museums.

Over the years, Spitz expanded its equipment line. The models B and C were developed in the over the next decade, followed by the A3P, which incorporated automatic planetary, solar, and lunar motions. The company sold hundreds of A3P instruments, and more than 300 are still in operation to date. In 1962, Spitz introduced the world's first planetarium instrument that was not strictly tied to an earth-bound point of view. It was called

Figure 3. A Spitz Model A. (Courtesy the Stanback Planetarium, South Carolina State University)

the Space Transit Planetarium and was followed in the 1970s by the model 512 and the Space Voyager in the 1980s. During the last decade of the 20th century and the first years of the new century, Spitz (now called Spitz, Incorporated) developed full- and partial-dome video projection systems for large theaters, called ImmersaVision and Electric Sky respectively, and the SciDome for smaller theaters. These are relatively high-end systems that pump computer-generated stars and video content onto a dome using video projection technology.

The expansion of planetarium theaters to hold all this technology was spurred in large part by advances in astronomy and space science beginning in the late 1950s. In the United States, the so-called "Space Race" in particular highlighted the need for advanced techniques in science and mathematics education. By the 1970s, more than 500 new planetarium facilities had been built in the United States. At the same time the US was beginning its headlong rush into space, Japanese industrialist, Seizo Goto directed his telescope making company to produce a planetarium instrument. GOTO's product line has grown to include both traditional (opto-mechanical) and video projection systems, with instruments which have since been installed all over the world. Also in Japan, the Minolta Company, part of the Minolta Group has been a player in the planetarium instrument field since the 1960s. Like their contemporaries, they began with opto-mechanical devices and have also moved into the video projection market with an affordable small-dome offering called MediaGlobe.

During this period when the other companies were developing their equipment and enlarging their customer base, Carl Zeiss did not sit still. Both Zeiss companies continued to create new projectors independent of each other, endowing the opto-mechanical planetarium with every engineering and optical advance. The most familiar Zeiss projectors, to US audiences anyway, were the later models – the Mark VI, which were installed in New York, Boulder, Colorado, Chicago, and other large cities. In 1976 a Mark VI was given to the National Air and Space Museum in Washington, DC as a gift to mark the occasion of the nation's Bicentennial celebrations. The Zeiss Jena counterparts began showing up in the former Soviet Union, East Germany and other countries where East German products were available.

Throughout the decades of the 1980s and 1990s, most of the major opto-mechanical instrument makers devoted a great deal of research and development into designs to maximize star field beauty. Nearly every company explored brighter lamp possibilities, and some began using fiber optic technology to deliver bright, realistic looking pinpoints of light to the dome.

The next revolution in planetaria was started largely by the Evans & Sutherland company of Salt Lake City, Utah. In 1983 they unveiled the Digistar 1, the world's first digital planetarium instrument. The early version of this instrument utilized a VAX computer with data stored on some 6,800 stars. A graphics processor displayed the data through a high-intensity cathode-ray tube (CRT) outfitted with a fisheye-lens projection system. In principle, this system would be completely familiar to today's users of computer-based "planetarium" and astronomy software packages like TheSky, Starry Night, SkyMap and others. The cost of the system lay largely in the VAX computer. The early skies were not very bright. This problem, coupled with the expense of the system limited the Digistar's reach in the market. With the advent of brighter projection systems and more powerful computers, however, the cost came down and today Digistar is aiming at the space theater and planetarium market with its Digistar III and Digistar Junior products.

In modern theaters, planetarium projectors aren't the only visual attraction under the dome – although their star projection capabilities are still the main drivers in programming philosophy. In the 1970s, planetarium producers and lecturers began adding supplementary audiovisual projection systems to add spice to their lectures and shows. These were an outgrowth of the single slide projectors and overhead projectors that speakers once used in the course of a lecture. Ultimately, systems of linked slide projectors called panorama and all-sky systems were developed to throw images onto the dome that would immerse the audience in a scene. Sound systems brought music and sound effects to the theater as well. Such audio-visual advances allowed planetarium show programmers to flesh out their lectures and even to plot out complex stories and presentations in much the same way that other multimedia and movie producers were doing. These technologies accelerated the production quality of planetarium shows while allowing producers to focus on the discoveries being made by astronomers in near real time. To present these programs effectively, however, all of the projection systems must be synchronized to the soundtrack. Early control systems were tone-based, relying on signal "beeps" on an unheard tape track to advance the projectors and in some cases control the star projector. Today computers do the controlling, using specialized systems to keep all the visual content in synchronization with the soundtrack.

In addition to the all-sky and panorama projection systems, planetarium producers began building special effects projectors to depict moving spacecraft, stellar explosions, and a host of other imagery in their shows. The rise of affordable video projection technology in the dome instantly enlarged the amount of visual content available to show presenters. The first facility to combine video with still imagery and planetarium stars was the Armagh Planetarium in Northern Ireland in the mid-1980s. It was a big first step, but many technical problems had to be worked out. Video projection systems are not specifically made for the extremely dark environments found in the dome, and they can easily "wash out" the stars and other projected images.

A number of companies, among them Sky-Skan, Incorporated (which also produces special effects and control systems), Evans & Sutherland, Zeiss, Minolta, GOTO, and Spitz Inc. have worked to improve image quality in these video projection units. The result has been a new explosion in planetarium projection systems termed "full-dome video." The first installations of these systems were the SkyVision system at the Burke Baker planetarium in Houston, Texas in 1998 and the Evans & Sutherland StarRider at Chicago's Adler Planetarium in 1999. Since that time, Silicon Graphics (a relative latecomer to the planetarium scene) designed and created

Figure 4. The Wichita, Kansas Cyberdome star theater. Equipped with an Evans & Sutherland StarRider and a Digistar planetarium projection system. (Credit: Exploration Place)

a video system for the newly renovated Hayden Planetarium at the Rose Center in New York City. It functions in tandem with Carl Zeiss's Mark IX star projector. Other installations of full-dome video include (but are not limited to) Exploration Place in Wichita, Kansas, Beijing Planetarium (China), and the National Air and Space Museum's Einstein Planetarium in Washington, DC.

4. The Portable Phenomenon

All of the instruments described above are installed almost exclusively in stand-alone theaters. In the second half of the 20th century a market arose for portable, table-top planetarium projectors. The most notable early example was the Spitz Junior. It could easily be set up in a classroom and was particularly attractive to teachers who had no planetarium nearby for their classes to visit. Today the market for portable planetaria is characterized by the Starlab instrument pioneered by Learning Technologies, Inc. (Somerville, Massachusetts), RS Automation's Cosmodyssée (France), and others. They are purchased and used mainly by schools, museums, and universities. A number of entrepreneurs have set up traveling planetarium businesses, offering programs to fit nearly every need to places where no facilities are available. In addition, these smaller movable domes can do double duty at science fairs, exhibitions, expositions, and other non-permanent public events. They are limited to use in smaller domes, and the typical traveling Starlab holds only a few dozen people comfortably.

5. The World of the Planetarium

Over the years, the term "planetarium" has come to be used as the name for the optical instrument that projects the motions and relative positions of the stars, planets, moon, and other sky objects. The name is also used to describe the room or building in which the projector is housed. For the rest of this chapter, we will use the term as it applies to the theater and planetarium instrument together as a place where planetarium presentations are made. The people who run planetarium facilities are known variously as planetarians or planetarium teachers/lecturers. There are several thousand planetarians in the worldwide community. Beginning in the 1960s and 1970s, several groups of planetarium educators came together to form alliances. Today the largest worldwide association of planetarians is the International Planetarium Society (IPS). It is something of a confederation of smaller groups like the Nordic Planetarium Network, the European-Mediterranean Planetarium Association, the Association of French-Speaking Planetariums, the assorted United States regional groups, and so on.

Most groups of planetarians meet every year while the IPS members meet bi-annually. The earliest meetings were sometimes small coffee-klatsch type gatherings that lasted a day or so and featured informal discussions about educational approaches to teaching astronomy. Gradually, as presentations became more complex, these meetings evolved into weeklong events. Today's planetarium meetings feature programs shown cheek-by-jowl with paper sessions and equipment demonstrations. The membership of IPS (and most smaller groups) includes teachers, writers, producers, lecturers, vendors (of equipment, shows, slides, books, etc.), and administrators.

Throughout the first part of the 20th century, there were only a few hundred planetarium facilities around the world. As discussed earlier, advances in automated assembly of planetarium instruments, coupled with less-expensive ways of creating opto-mechanical systems contributed to a rapid rise in the numbers and types of facilities. The inventors of the first instruments would be amazed at the population of facilities out there in the first years of the 21st century.

As of August 2002, there are 2,959 planetariums of all sizes, theater configurations, and philosophies around the world. 1,503 of those facilities are in the United States and 1,456 are distributed across 89 other countries. Japan has more than 300 planetarium theaters, while Germany has more than 100. The most recent estimates on planetarium attendance are not precise, but even a so-called "ballpark figure" has nearly 90 million people each year visiting a planetarium.

Planetarium facilities today can be sorted into the following categories: school theaters, university/college facilities, museum/public planetaria, and unclassified. About 2/3 of the facilities are school-based, with the rest being divided between university and public facilities[1].

From these numbers it is easy to see the influence that educational priorities have had in the construction and design of planetarium theaters. Some of these facilities are actually located in square rooms (former classrooms), or in additions to school buildings. They usually do not have as many seats as public facilities do, and some of the older ones are equipped with little beyond the star projector and a few auxiliary slide projectors. Educational requirements also affect the production of content for school planetarium facilities. Their offerings have always been very much guided by curriculum needs, and more recently educational standards. The presentations range from lectures and multi-media star shows to participatory lessons where students interact with the planetarium presenter/teacher. While many school districts and higher education facilities are in stand-alone theaters and some can offer fairly complex shows, some are either portable or have portable domes in addition to their regular theaters.

Theater design and programming philosophy are somewhat different in public facilities. The theaters themselves may be in stand-alone buildings, or as part of a larger science center/museum complex. Many of them seat up to several hundred patrons, and in some theaters the emphasis is on moving large numbers of patrons through the theater in the course of a day.

[1]Data on planetarium facilities and attendance supplied by Loch Ness Productions.

The museum/science-center shows are often very lavishly produced multimedia or full-dome video presentations. They are written to appeal to a so-called "general public" level. In addition to the main feature, many public facilities also offer a "library" of school shows for use by visiting groups. On the purely entertainment side, some planetaria have offered laser-light shows and other experimental types of shows. In addition, these facilities will schedule a regular "What's up" or "Sky Tonight" show to acquaint those attendees who are interested in finding their way around the night sky. Public astronomy lectures on timely topics in astronomy and space science are usually offered on a selective basis.

Beyond these functions, many public facilities (and a smaller number of college and school theaters) are used as presentation spaces for concerts, plays, fund-raising gatherings, weddings, funerals, and other events not necessarily associated with astronomy education. Often the very largest museum facilities are offered for social functions – often advertised as a "rent the planetarium for your next function" approach to making money for the museum. This is not much different from the way that other large spaces in public buildings and theaters are available for use by fee-paying organizations. While important to individual facilities for their fundraising power, these activities are not typical of the function of the planetarium. Rather, the fact that they happen at all is a tribute to the interesting ambience that a planetarium theater can provide for a social function.

6. The Anatomy of a Planetarium Show

Over the past few decades, philosophies in show production have often polarized along the "party lines" of education and entertainment. For a while in the 1970s and 1980s, planetarium professionals (who have backgrounds ranging from education to media production) debated the relative merits of educational programming and shows produced to be solely entertaining. Similar controversies erupted over whether live lectures were better than taped programs.

As discussed here, today's show offerings cover a wide range of presentation styles and levels, and most producers make an effort to education AND entertain their audiences – no matter what venue the planetarium is in. It is difficult to define a "typical" planetarium show since there are so many ways that planetarians utilize their theaters. Still, it is useful to have a brief discussion about how the majority of pre-recorded presentations are produced, since their creation is largely affected by the hardware through which it will be projected.

At its heart, a planetarium show is a mix of visuals and sound, utilizing some sort of projection system in conjunction with the starfield projector.

These shows approach the audience in both the cognitive and affective domains, offering a mix of science, art, and a greater appreciation of the beauty of the universe.

Like other multimedia presentations, a planetarium show begins as a script and storyboard. The script is researched, written, frequently "vetted" by astronomy professionals, and eventually narrated. The narration is mated with music and sound effects. While the script is in production, visual materials are created. In a "traditional" planetarium theater equipped with slide projector systems, these visuals are shown as 35mm slides. Some theaters have also used 16mm film material and special effects projectors (such things as cloud projectors, zooms, slews, and others) to supplement the inventory of images. [Note: the film projectors are not the same as Omnimax^{TM} projection systems, which project IMAX^{TM} movies on the dome.] Visually dense shows may have several hundred images or video sequences.

There is no set length for a planetarium show, and preferences have changed over the years. In a school facility, the class period length may well determine the duration of a presentation. In public facilities, there are different concerns. For example, a given museum may wish to have audiences to through the planetarium every 30 minutes. Another facility may favor 40-minute shows. In the 1970s and 1980s, programs can and did last up to an hour or more! These days, the trend seems to be for shorter programs.

One factor influencing this direction is the cost of show production. Slide-tape presentations are much less costly than video animation sequences (at least for the visual production part of the show). While some facilities prefer to create their own programs, many also rely on readily available programming created by a variety of producers. Since most are geared for slide-tape presentation, it's easier and cheaper to buy these shows and install them. The current costs to produce a slide-tape program in today's market can run anywhere from a few hundred dollars (US) to upwards of $100,000, although shows at the higher end may include some video clips as well. Distribution packages can be as little as $200.00 (or free in some limited cases) to around $10,000 per show package. Each package contains visual materials (usually slides), a soundtrack on CD or digital tape, and a script/production notes package. Planetarium producer then take the packages and install the various pieces and parts into whatever projection systems exist in their theaters.

That model changes when the medium is full-dome video. First of all, production costs are significantly high enough that custom planetarium shows utilizing video clips or full-dome content can run up to around a quarter of a million dollars. Distributed show packages can be as little as a few thousand dollars to $50,000 or more.

The distribution is a bit trickier because each technology takes its data in a different way. Some systems read the show data directly off of a CD-ROM (or series of CDs). Others read from an array of hard drives, while still others are programmed to crunch the show data and project it on the dome in real time. Once a show is installed on the medium of choice, however, it runs at the push of a button without the need for the panoply of auxiliary systems still prevalent in many of the world's facilities.

Show production packages are still relatively rare for the full-dome video community. Currently a majority of these types of shows are available from the museum facilities that commissioned them. A few companies are gearing up to produce shows for general distribution, and sales models are still being worked out. Still, in these early days of full-dome video, planetarians who are buying into new technologies like the Digistar III, the Sky-Skan Skyvision system, the Spitz SciDome, or the Minolta MediaGlobe are facing a rather limited set of program choices. This will change as more producers convert their existing inventory to full-dome and begin to create new programs. And eventually those who buy the new hardware will come up to speed on producing for their new toys. Still, for facilities used to paying much less for slide production, video production costs are an important driver in determining the length of a show.

7. Trends at the beginning of the 21st Century

The planetarium and its community have come a long way since Walter Bauersfeld first turned his engineers loose on the problem of projecting stars onto a dome. The evolution of the planetarium instrument from an opto-mechanical clockwork wonder to a computer-based immersive system continues in parallel with changes in theater design and personnel. The advent of full-dome video projection systems is having a marked effect on the profession. At the 2002 International Planetarium Society meeting in Wichita, Kansas, attendees had a chance to sample most of the up-and-coming technologies in planetarium instrumentation and theater equipment. For many it was their first chance to find out just what will be required of them in the brave new world of show production. Certainly the revolution in projector styles will not sweep all of the old systems out of theaters overnight. The majority of older projectors in place in star theaters around the world is in fine shape and is expected to run well for quite some time. However, the auxiliary projection systems that so many facilities rely upon for the immersive environment may well go obsolete – particularly if such things as slide mounts and specialty films needed to produce slide-tape shows go out of availability.

When they do, and facilities start buying more of the full-dome video installations, planetarians themselves will face a rather exciting and steep production learning curve. Already some of the older museum facilities are following the lead of the Forum der Technik in Munich, New York's Hayden and Chicago's Adler Planetariums in mating the latest technologies together in fully immersive theater environments. Currently the venerable Griffith Observatory in Los Angeles is working on a much-needed renovation utilizing some of the newest technology to help them present their traditional live lecture fare. The Denver Museum of Nature and Science is working on even more cutting edge technology to bring their old Gates Planetarium into the 21st century. Even the smaller domes are not immune to this change. As of this writing, Minolta is preparing to install the first generation of its small-dome video MediaGlobe systems in the United States. Learning Technologies has updated its StarLab projector with brighter lamps and new overlay cylinders that extend the portable system's use.

Despite all these advances and shifts, some things do not change for planetaria and their users. The need for accurate sky representations is as important as it ever was – indeed, in these days of increasing light pollution and less access to truly dark skies, planetarium theaters are fulfilling that need that humans have to experience pristine skies and learn more about the stars. In that regard, they have never left the roots set by their inventors. Their growth into a unique medium is a tribute to their essential link between the astronomy community and the general public. Planetariums will always have a place in the world's educational and cultural institutions because of their ability to combine an appreciation of the night sky with the latest in astronomy research. There is no doubt that the coming century of technological evolution in the planetarium will echo and mirror the technological change in astronomy research. Ultimately, this evolution will prove every bit as exciting as the past 100 years have been for such a fascinating profession.

Acknowledgements

Special thanks to: Jordan Marché, Mark C. Petersen (Loch Ness Productions), Laura Misajet (Seiler Instruments/Carl Zeiss), Ken Miller (GOTO Optical Mfg. Co.), Virginia Savage (Sky-Skan, Inc.).

References

1. Abbatantuono, B.P. Armand Spitz – Seller of Stars, Courtesy of Griffith Observatory web site: http://www.griffithobs.org/IP
2. Chartrand, M.R. 1973, A Fifty Year Anniversary of a Two Thousand Year Dream

(The History of the Planetarium), *The Planetarian*, September issue
3. Gutsch, W.A. & Manning, J.G. 1998, The Current Role of Planetariums in Astronomy Education, in *New Trends in Astronomy Teaching*, Cambridge University Press.
4. Hagar, Ch.F. 1998, Window to the Universe, Carl Zeiss, Inc.
5. International Planetarium Society resources web page: `http://www.ips-planetarium.org/ips-welcome.html`
6. Petersen, M.C. 2002, The Loch Ness Planetarium Compendium, Loch Ness Productions.
7. Steve's Planetarium Resources and Collectors Page web page: `http://www.pielock.com/ppr-collector.html`
8. The History of the Planetarium, Courtesy of the Morrison Planetarium web site: `http://www.calacademy.org/planetarium/about.html#brief`

THE CHANGING ROLE OF THE IAU IN PROVIDING AND ORGANIZING INFORMATION

A.H. BATTEN
National Research Council of Canada
Herzberg Institute of Astrophysics
Dominion Astrophysical Observatory
5071 West Saanich Road
Victoria, B.C., Canada, V9E 2E7
alan.batten@hia.nrc.ca

AND

D. MCNALLY
Department of Physical Sciences
University of Hertfordshire
College Lane
Hatfield AL10 9AB, UK
dmn@star.herts.ac.uk

Abstract. We discuss the way in which the IAU has organized information and supplied it to the astronomical community (with special reference to the Telegram Bureau), to students worldwide, to other international unions, to governments and inter-governmental organizations and to the general public.

1. Introduction

The International Astronomical Union (IAU), one of the first international scientific unions, was founded at conferences in London and Brussels in 1918 October and 1919 July, respectively, and its first General Assembly was held in 1922. From the beginning, the provision and organization of astronomical information has been one of the IAU's *raisons d'être*. Much of that information concerns only the relatively small groups who make up the IAU's (currently) 37 Commissions. To discuss the specialized information provided by each Commission is beyond the scope of this brief survey,

A. Heck (ed.), Information Handling in Astronomy – Historical Vistas, 249-266.

and would require a volume of its own of only limited interest. Those wishing to know what the various Commissions are doing are referred to the most recent "A" volume of the *IAU Transactions* (currently Vol. XXIVA, 1999). The IAU as a whole, however, is concerned with the organization and dissemination of information of more general interest and it is on these functions that we shall focus. We shall look at the way in which the IAU handles information of value to the astronomical community (especially the Union's own members), to students, to other international unions, to governments and inter-governmental organizations, and to the public.

Throughout its eighty-year history, the tasks and methods of operation of the IAU have changed. The IAU and its Commissions rapidly adopted e-mail for internal communications, whenever possible, and increasingly disseminate information electronically, on web pages, relying less on the distribution of masses of paper. A worldwide organization, however, cannot eliminate printed documents overnight; in many countries people still depend on the printed page. We shall try to look at the changes that have taken place and are still going on, as well as at the changing substance of the information disseminated and the new types of information required.

2. Information for the Astronomical Community

Astronomers, more than any other scientists, need to cooperate internationally: no one person on the Earth can observe the whole sky, or monitor a single object continuously. Observations of unpredictable and transient events (novae, new comets *etc.*) must be communicated promptly; those of predictable but rare events (total solar eclipses or, formerly, transits of Venus) must be coordinated in advance. Surveys of the whole sky demand worldwide collaboration. Above all, regulation of time services must be coordinated globally.

For these reasons, even before the IAU was founded, astronomers were working together, through International Conferences on the Meridian or on Time, through such bodies as the Royal Astronomical Society and the *Astronomische Gesellschaft*, or through *ad hoc* collaborations between individuals – *e.g.* O.W. Struve and Newcomb collaborated in arranging observations of transits of Venus (Batten 1988). Not all these initiatives were successful – the *Carte du Ciel* [1], for example, was flawed and never completed; even so, the astrographic telescopes built for it have uses that their designers probably never envisaged (Fierro 2001).

[1]See the chapter by D.H.P. Jones in this volume. (Ed.)

2.1. IAU PUBLICATIONS

The IAU publishes five periodic series – *Transactions* (A: *Reports on Astronomy* and B: *Proceedings of the General Assembly*), a Symposium series, *Highlights of Astronomy* (the scientific proceedings at General Assemblies) and an *Information Bulletin (IB)*. Proceedings of IAU Colloquia are also normally published but not by the IAU itself.

A recent trend is for Colloquia to be published by the same publisher (Astronomical Society of the Pacific – ASP) as the IAU Symposia, within the framework of the ASP Conference Series. The *IAU Transactions* have been published since the first General Assembly in 1922, splitting into A and B in 1961, and are published triennially. The IAU Symposium series began in 1953 (published in 1955) and by the end of 2002 the IAU will have supported some 215 symposia. *Highlights of Astronomy* records briefly the scientific proceedings at General Assemblies since the XIII General Assembly in 1967. The *IB* began in June 1959. Issued now twice each year, it will reach No. 92 in 2003 January (see Sect. 2.5).

By the end of 2002 the IAU will have supported 190 colloquia since they began in 1959. Symposia have been held at the rate of just under 4.4/year, while colloquia have been held at just over that figure. The foregoing are the major periodical publications of the IAU and its greatest service to the dissemination of astronomical science within the community.

The IAU has also published the proceedings of other symposia/colloquia (*e.g.* the early Cosmic Dynamics meetings) conjointly with other Unions and the proceedings of its regional meetings (European – eleven until 1989 when the European Astronomical Society took over, Asian-Pacific (eight meetings to the end of 2002) and Latin-American (also eight meetings to the end of 2002), thus making a significant contribution to publishing current astronomical research.

The IAU has also undertaken other significant publications; some for only a short period, to serve a particular need, others almost for the life of the Union or even longer (see next subsection). Examples of publications produced with IAU support can be found in the following references: Delporte (1930a&b), Arcetri Observatory (1926-1939) and Cayrel *et al.* (1980).

The IAU has also supported the publication of the *General Catalogue of Variable Stars* (Kholopov *et al.* 1985-95) since 1959, and various Commission newsletters (especially the *Newsletter* of Commission 46 since 1971) and bibliographies (*e.g.* Commission 41's Bibliography of books and papers on the history of astronomy).

Thus the IAU has a record of supporting the publication of information of value to astronomical science when that information could not be disseminated in any other way. It is a substantial record of support – often achieved at minimal cost, if also subsidized by individual observatories and institutions.

2.2. ASTRONOMICAL TELEGRAMS

We shall discuss astronomical telegrams in more detail than many other projects because they illustrate a paradigm of the way the IAU should serve the worldwide astronomical community. The Central Bureau for Astronomical Telegrams was established by the *Astronomische Gesellschaft*, following its 1881 meeting in Strasbourg and set up in Kiel, Germany, at the editorial office of *Astronomische Nachrichten*, the editor of which was to be the Bureau's Director. Since European astronomers, at least, were accustomed to communicating their discoveries to that journal, this arrangement was obviously sensible: they now had the opportunity to use telegraphy. Institutions wishing to receive the telegrams subscribed to cover their cost.

The arrangement worked well until the outbreak of World War I, when astronomers in countries in conflict with the Central Powers could not communicate with Kiel. By mid-September 1914, it became clear that interim arrangements were needed. Even then, one aspect of the Bureau's work had become the assigning of priority for discoveries and there were four competing independent claims, from four different continents, for the discovery of Comet 1914*e* (Lecointe 1922). Early in November, Ellis Strömgren, in neutral Denmark, offered, with the agreement of Kobold, then editor of *Astronomische Nachrichten*, to receive and transmit astronomical despatches. Despite his high personal reputation, however, Strömgren was perceived by some as having too close a relationship with the Kiel Bureau: many astronomers in the allied countries refused to use his services. Baillaud, in Paris, coordinated astronomical despatches among the Allies.

Blaauw (1994) has described how wartime animosities affected the IAU's early history and the above summary shows that the telegram service was particularly disturbed. Putting the service on a peace-time footing was one of the IAU's first tasks. Commission 6 of the IAU was formed in 1919 to supervise the Bureau's activities. Lecointe offered to house the Bureau at the Royal Observatory in Brussels; it opened there on 1920 January 1. The former Central Powers, however, being excluded from the IAU, continued to use the services of Strömgren in Copenhagen. This arrangment was clearly unsatisfactory, even though Strömgren and Lecointe cooperated and shared mailing lists. In 1922, at the first General Assembly, Lecointe announced that the Observatory in Brussels wished to be relieved of the Bureau not later than the end of that year; the Bureau moved to Copenhagen in October of that year (Fowler 1922), remaining there until it moved to its present home, the Smithsonian Observatory, on 1965 January 1 (Gingerich 1966).

Strömgren remained Director until his death in 1947, followed by Vinter Hansen until her own death in 1958, and by Thernöe, who remained in office until the transfer to Cambridge, Massachusetts.

According to Lecointe, of 38 institutions subscribing to the telegrams when the Bureau was transferred from Brussels, only five were not in Europe; three were in Africa (although he counted the observatory in Algiers as European!), one in Australia (Melbourne), and one in America (Harvard). Only one American institute was included because Harvard served as a distribution centre for the entire continent and, from 1926, issued the *Harvard Announcement Cards*. The absence of Russian institutes is not surprising, given the turmoil in that country then (Vladivostok subscribed soon afterwards) but the complete absence of any Asiatic country perhaps was. By the time of the transfer to the Smithsonian Observatory, subscribers had the option of receiving telegrams or mailed *Circulars*; 85 institutes subscribed to both and 242 to *Circulars* only (Thernöe 1966). The transfer to Smithsonian nearly doubled the subscription list, since the *Harvard Announcement Cards* were discontinued and North American observatories received the *Circulars* directly.

Keeping costs low was a constant concern of Directors, in Brussels, Copenhagen and Cambridge (Mass.). A telegraphic code reduced the cost of individual telegrams by condensing information; updates to the code are regularly found in early volumes of the *IAU Transactions*. A fully modernized version was printed, after the transfer to the Smithsonian Observatory, in the *Astronomers' Handbook* (Gingerich 1966). There were active debates at several General Assemblies about the nature of the information to be encoded and tension is evident between the needs for complete descriptions (especially of comets) and the low costs. Sending only a telegram to only one centre on other continents also reduced costs; at times, other organizations have helped to pay for local distribution from such centres. The constant growth of the originally self-supporting service led, as early as 1928, to a request for an IAU subvention for the Bureau; a subvention that has been a feature of the IAU accounts ever since.

World War II created further administrative and financial problems for the Bureau. The former were solved by the Observatories in Lund and Zürich, in neutral countries, which relayed telegrams from occupied Copenhagen to countries at war with the Axis powers. Because IAU funds could not be transferred into Denmark during the occupation, grants and loans were made from Danish foundations. After the War the Bureau carried a large debt for some years and post-war inflation considerably increased the cost of telegrams. Already in 1961, Commission 6 considered the use of TELEX by the Bureau (Buchar 1962), but telegrams were still working well and, neither then nor when the Bureau moved to the U.S.A., did a change appear desirable. The first Director of the Bureau in its American location was Owen Gingerich who, however, resigned shortly afterwards, in favour of Brian Marsden, who held the position until 2000, to be succeeded by the present Director, Daniel Green. The Director of the host institution, F.L. Whipple, often spoke for the Bureau at meetings of Commission 6. In 1970, he told the Commission that the telegram was a "dying institution" at least in the U.S. and serious thought was being given to the use of TELEX or "mailgrams" – the latter were cabled to the appropriate postmaster, who arranged local delivery (Whipple 1971). The Bureau was not then a 24-hour operation and a TELEX machine would at least enable incoming messages to be received at any time. In fact, the Bureau acquired two TWX machines in 1978. Changing financial and administrative arrangements between the host institution and the Bureau are also reported in this period.

Subscription numbers grew considerably in the 1970s but so did costs. The type of information submitted to the Bureau for the *Circulars* caused concern. Instead of just alerting astronomers to transient phenomena, *Circulars* were being used to communicate pulsar discoveries and, again, to claim priority. Increasing interest in minor planets was also reflected in the content of the *Circulars*. Some of this material was deflected into the *Minor Planet Circulars*, a process no doubt helped by the fact that the Director of the Bureau was also the Editor of the *Minor Planet Circulars*. The Bureau's Director was authorized to edit, or even to refuse, material that he considered should be more appropriately published elsewhere and, in 1979, "line charges" were introduced for publication in the *Circulars*.

Early in the 1980s, Commission 6 again discussed "electrical communication" (Hers 1982); information was now sent by TELEX to distribution centres in Moscow and the southern hemisphere. The discoveries of Comet IRAS-Araki-Alcock in 1985 and the outburst of Supernova 1987A broke all records for the Bureau's activity, but delays in printing and delivery led to a decline in the number of subscriptions – which had previously increased steadily. These facts precipitated changes in working methods at the Bureau, which began to use the computer network SPAN. Telegrams were still needed outside North America, Western Europe, Japan, Australia, New Zealand and South Africa, but within those countries, 50 percent of the subscribers could be reached by SPAN or BITNET. The *Circulars*, wrote the Director (Marsden 1990), had changed from "a printed publication with an electronic version" to an "electronic publication with a printed version". By 1991, 25 percent of the subscribers world-wide were electronic and telegrams were discontinued two years later, remaining only in the names of the Commission and Bureau.

The rapid growth of the World-Wide Web took everyone by surprise, including possibly its originators, and it is fair to say that the Bureau is still adjusting. Anyone can now see the titles of new *Circulars* on the IAU web page, but only subscribers can read them. Restricting a web page to subscribers is not difficult but preventing others from posting freely what they find there is. Recent reports of the Bureau (West and Marsden 1997) show concern about this sort of "piracy", which has affected the number of subscriptions, now considerably below their maximum number in the 1970s and 1980s. The implications are serious; what was once a small additional chore for the editor of *Astronomische Nachrichten* has become a round-the-clock operation employing three people, only one of whose salaries is paid by the host institute. Despite that institute's subsidies, the IAU subvention and occasional subsidies from other organizations, the costs of providing the service astronomers want must be largely met by the users. In the face of pressures and difficulties, the Bureau has maintained a very high standard; false alarms are inevitably received but very few, perhaps only one serious one, have been disseminated. Surely the astronomical community will not allow a service of such demonstrated utility to lapse, but the story of how the IAU handles this information is still developing. At least we know that the IAU is not alone in trying to balance the desire to disseminate information as widely as possible with the need to recover the costs of gathering and organizing it.

2.3. MINOR PLANET CIRCULARS

Another major IAU contribution to the organization and dissemination of data has been the *Minor Planet Circulars.* Although more closely the concern of a regular Commission (20) than the Telegram Bureau, these circulars have become important enough to deserve special mention. The *IAU Circulars* and the *Minor Planet Circulars* overlap, to some extent, since one of the original purposes of the former was to alert observers to newly discovered minor planets so that an orbit could be determined and the planet recovered after conjunction with the Sun. The *Minor Planet Circulars* have a larger scope, recording observations and ephemerides, assigning final numbers and, eventually, listing names and citations. Again, the fact that for a long time the two publications had the same editor has, in several ways, enhanced cooperation between the Central Bureau of Astronomical Telegrams and Commission 20, reducing duplication to a minimum.

Once again, wartime history impinged on the IAU and led to a permanent innovation. Before World War II circulars about minor planets were published by the *Astronomisches Rechen-Institut* in Heidelberg, which could not continue that work in the post-war ruin of Europe. Paul Herget offered to issue *Minor Planet Circulars* from Cincinnati Observatory (Delporte 1950). The earliest circulars, single typed sheets, like the telegrams, were to be self-supporting, paid for by the subscriptions of the recipients. Indeed, the only cost, at first, was air-mail charges for Observatories outside the U.S.A. The enterprise grew; reports of Commission 20 often contained statements to the effect that "during the past triennium twice as many circulars were issued as in the previous one". The circulars are still single sheets, but the sheets are closely printed and distributed more frequently.

Major reasons for the increasing interest in minor planets have been the desire to know more about the numbers of objects in orbits that intersect with that of the Earth and study of Kuiper-Belt Objects. Cincinnati Observatory set up a Minor Planet Center to deal with the *Circulars* and to act as a clearing-house for minor-planet information and as a computation centre for orbits. At the General Assembly of the IAU held in Moscow in 1958, Herget announced by letter (Herget 1960) that the Center had now acquired an IBM 650 computer and all the information it needed from observers, in order to compute preliminary orbits and ephemerides, was the Julian Date of observation, the observed right ascension and declination and the name of the observatory. The Center remained in Cincinnati until Herget's retirement, when it moved to the Smithsonian Observatory, on 1978 July 1, under the direction of Brian Marsden. As with the *IAU Circulars* themselves, there is an evident trend toward electronic transmission of the *Minor Planet Circulars* whenever possible.

2.4. BIBLIOGRAPHY

The preservation of information in books or electronic storage is clearly a concern of the IAU although, unfortunately, one to which rather few working astronomers pay attention. Another of the IAU's old commissions, Commission 5, worked in this area. Originally named *"Commission des analyses de travaux et de bibliographie"* (a name that never seems to have been given a precise English equivalent) it was renamed *"Documentation"* at the XIII General Assembly in 1967, when it became a committee of the Executive Committee, and again *"Documentation and Astronomical Data"* at the XVII General Assembly in 1979. These name changes were not purely cosmetic, but signified a continuous widening of the Commission's sphere of interest. Currently it comprises five Working Groups: on astronomical data, libraries, designations, FITS (Flexible Image Transport System) and virtual observatories, data centres and nedtworks. In this way, the Commission tries to oversee, in the best intersts of working astronomers, the developments of electronic storage, retrieval and transmission of data.

In line with its original name and purpose, the Commission maintains liaison with libraries and has tried, with mixed success, to persuade astronomers to adopt standard, unequivocal terminology. Although the IAU

Figure 1. Three past officers of IAU Comm. 5 Working Group on Astronomical Data (B. Hauck, G. Westerhout & A. Heck) in Viking hands at York during the 24th IAU General Assembly (Manchester, August 2000).

issues no journal of its own, as we have seen, publication of symposia proceedings is one of its major activities, so it has a legitimate interest in setting publication standards. A large part of the original *Astronomers' Handbook* (Pecker 1966), which has been up-dated by Commission 5, contains proposed abbreviations for references. While these could be mandatory only for IAU publications, several major journals adopted the abbreviations which, never particularly popular with the average working astronomer, were clearly designed so that librarians who were not specialists in the astronomical literature could reconstruct titles unambiguoulsy.

Thus our familiar *Ap. J.* became *Astrophys. J.* – an elaboration that seemed unnecessary to many. The distinction between *Astrophys.* (for "astrophysics") and *astrophys.* (for "astrophysical") was a refinement that made less sense to anglophone authors than to francophone ones. Two factors eventually led to the abandonment of this system: first, in the early days of preparing camera-ready copy not all the italics and bold-face of a traditional citation could be reproduced, and second, the pressures of space in journals. We have returned to very condensed abbreviations and, ironically since different type faces can be produced at the word-processor, eliminated varied type faces from many journals. Perhaps the pendulum has swung too far toward abbreviated references that are nearly incomprehensible, but the episode illustrates an area in which the IAU was less successful in influencing the practice of the astronomical community. We might suggest that, when the IAU is responding to the wishes of the community, it usually does an excellent job of coordinating the organization and presentation of data but that, when it tries to impose something from the top, it is often much less successful.

This same lesson may have been taught by the attempt in the *Astronomers' Handbook* to define a uniform transliteration from the Cyrillic alphabet. Here again, the obvious aim was to enable anyone who did not know astronomy to recreate the Russian word being represented. Mostly, the system was used only for proper names and few of us could bring ourselves to write the name of our colleague Ambartsumian (himself an IAU President) as "Ambarcumjan"!

2.5. THE UNION'S INTERNAL INFORMATION

The IAU's membership, initially just-over 200 in 19 countries, has grown to nearly 8,600 in 66 countries (plus about another hundred in 18 non-adhering countries) at the General Assembly of 2000. Between Assemblies, the IAU is administered by an Executive Committee of about a dozen members, in particular, by three officers: President, General Secretary and Assistant General Secretary. The need to keep individual and national members informed of the IAU's affairs clearly has increased many-fold and will probably continue to do so. Until the late 1950s, the General Secretary sent out "circulars" to the members, as circumstances required, but in 1959 these communications were regularized, as mentioned in Sect. 2.1, by the *Information Bulletin*, so-called to avoid confusion with the *IAU Circulars*, The first *IB*, of eleven printed pages, appeared in 1959 June; the most recent, at the time of writing (No. 90, 2002 January) had 42 pages.

Sometimes, just before a General Assembly, they can be even larger. Host countries used to distribute a *Preliminary Announcement* and a *Preliminary Programme* of the Assembly to each member, and to provide a *Final Programme* for each participant on arrival. As one of us (AHB) can testify from experience, by 1979 printing and mailing costs of these distributions were a major part of the host country's budget for an Assembly. From 1982, much of this information appeared in the *IB*, printed and distributed at the IAU's expense rather than the host country's. For the most recent

Assemblies, the information has also been available electronically. The IAU set up its own web page in the 1990s which now has a permanent home at *www.iau.org*, on which both the *IB* and the membership list appear. Electronic distribution is thus becoming more important in IAU administrative affairs, as it is in the scientific, although it may never entirely supersede printed versions.

3. Information for Students

Provision and organization of information inevitably includes passing information to the next generation, which is a part of teaching. The IAU's principal role has been the coordination of research; there is not much about teaching in early volumes of the *Transactions*. Concern for students began to appear, slowly at first, after the end of World War II. The first sign was the founding of Commission 38 (Exchange of Astronomers) by the Executive Committee at its first post-war meeting in Copenhagen in 1946 (Oort 1950). F.J.M. Stratton was appointed the first Commission President and remained in charge for nine years. In those years, UNESCO provided much of the funding for the exchange of astronomers between countries. Early lists of grants suggest that the chief purpose was to encourage visits to and from the war-devastated countries of Europe, but this was not explicitly stated and, even in the first list, two grants for Chinese astronomers are recorded. Graduate students were eligible for awards and a sub-commission was soon formed to deal with their applications. Subsequent lists of grants awarded have encouraged exchanges between all countries.

The next step was a meeting on the teaching of astronomy held during the XII General Assembly in 1964 (Minnaert 1966). Almost immediately (at the closing session of the General Assembly) the IAU approved the creation of Commission 46 (Teaching of Astronomy), which has become the focus for all IAU activities concerned with providing information to teachers and students. In 2000, at the XXIV General Assembly in Manchester, Commissions 38 and 46 and the Working Group for the Worldwide Development of Astronomy were combined in an enlarged Commission 46: Astronomy Education and Development.

Another strand was a six-week summer school held in Manchester in 1967, primarily for students from developing countries, run jointly by Z. Kopal and J. Kleczek, partly with UNESCO funding. Commission 46 speedily endorsed this initiative and its continuation, as International Schools for Young Astronomers, became one of the Commission's major activities. Kleczek remained in charge until 1991, followed by D.G. Wentzel and, in 1997, M. Gerbaldi. An ISYA is held almost every year and they have now been held in many different parts of the world. ISYA have been a major means of conveying modern astronomical information to students who live in countries where that is not always available; many graduates of ISYA have become notable astronomers in their own right.

A more ambitious project was the Visiting Lecturers' Programm (VLP), jointly discussed for several years by Commissions 38 and 46 before it was instituted at the XVIII (Patras) General Assembly in 1982 (Ferraz-Mello 1983). The aim was to bring in enough visiting lecturers to give a whole course in modern astronomy in countries where local resources were inadequate to the task. VLPs were set up in Paraguay and Peru. A VLP was proposed for Nigeria but could not be arranged; another was discussed for China, but that also did not materialize. Only the Paraguay VLP was completed. It has proved difficult to find people willing and able to spend the relatively long periods needed by the host country. Because of these problems, the VLP has been superseded by a Teaching for Astronomy Development programme (TAD), conceived by D.G. Wentzel. TAD also requires visits to the host country and still meets difficulties on that score, it can include bringing people from the host country to developed countries and the provision of aid by, for example, writing text-books for translation into the language of the host country. Considerable success has been achieved by TAD programmes in Viet Nam and Central America. One has also begun in Morocco and another is under consideration for The Philippines.

From the beginnings of Commission 46, members have discussed how far the Commission should be involved in education in primary and secondary schools, as opposed to universities. There is much need for help at the lower levels, especially in developing countries, but IAU resources are limited and must not be spread too thinly. The Commission has not ignored schoolteaching, but its efforts have mainly been confined to helping teachers. Beginning in Sydney, Australia, at the XV General Assembly in 1973, a day-long workshop (now traditional but originally the initiative of one of us, DMcN) has been held for schoolteachers from the host country or region, either just before or just after the General Assembly proper. Sometimes, similar workshops have been held in conjunction with IAU Regional Meetings, *e.g.* in Pune, India, in 1993. Thus the IAU has reached teachers in many different countries, introducing them to current ideas about teaching astronomy and enabling them to make contact with colleagues from the same country or region who may be facing problems similar to their own.

The growth in IAU activity designed to pass on information to younger generations may appear somewhat *ad hoc*, but it can also be seen as adaptation to needs as they are recognized, programmes being developed and modified in the light of experience. Teachers, even at school level, are being encouraged to use electronic data bases and some of the techniques of distance-learning, but the IAU itself still finds that personal contacts, in meetings, workshops and longer visits, are the most effective way of assisting the transfer of information from one generation to the next.

4. The IAU and Sister Scientific Unions

The IAU has collaborated with other Unions of the International Council of Science (new name for ICSU) family since its foundation and has played an active role in ICSU itself. Clearly some unions are closer than others to the scientific interests of the IAU, *e.g.* IUGG, URSI, IUPAP; but there is collaboration with other unions. Most recently the International Union for the History and Philosophy of Science (IUHPS) and IAU Commission 41 (History of Astronomy) have formed the InterUnion Commission for the History of Astronomy (Stephenson *et al.* 2002). A major partner is the ICSU Interdisciplinary Body, the Committee on Space Research (COSPAR) (established 1958) with whom the IAU has sponsored many joint scientific meetings (*e.g.* at the 2002 COSPAR Scientific Assembly the IAU cosponsored seven meetings and will cosponsor the second COSPAR/IAU Capacity Building Workshop in X-ray Astronomy in 2003 January). An IAU representative serves on the COSPAR Bureau. A recent collaborative venture was an IAU Commission 46 initiative, cosponsored by COSPAR, for a Special Workshop on Astronomical and Space Education held during the 1999 UNISPACE III in Vienna: as a result, the UN Committee on the Peaceful Uses of Outer Space (COPUOS) requested a project on Educational Capacity Building for Space to be led jointly by the IAU/COSPAR.

The IAU, IUGG and URSI sponsored the formation of the ICSU Interdisciplinary Body, the Federation of Astronomical and Geophysical Data Analysis Services (FAGS) in 1956. Of eleven FAGS Services, four are strongly astronomical – the International Earth Rotation Service (IERS) (jointly with IUGG, URSI), the *Quarterly Bulletin on Solar Activity*, the *Centre de Données astronomiques de Strasbourg* (CDS) (jointly with IUGG) and the Sunspot Index Data Centre. In 1960 IAU, URSI and COSPAR were the sponsors of an ICSU Interdisciplinary Body, the Scientific Committee on Frequency Allocations for Radio Astronomy and Space Services (IUCAF). IUCAF was recognised as the negotiating body for Radio Astronomy and the Space Sciences by the International Telecommunications Union (ITU – an intergovernmental body) in 1963 and has performed sterling services in maintaining radio-frequency bands for radio astronomy and the space sciences. IUCAF has all too often had to fight intrusion by transmitting services into bands designated by the ITU as radio-quiet for the use of the passive services. Management of the limited radio-frequency spectrum is difficult, demanding, time-consuming and is carried out by relatively few, but intensely dedicated, people on behalf of the passive services.

ICSU has encouraged collaboration between its Unions and Interdisciplinary Bodies. The IAU maintains relations with the Committee on Data for Science and Technology (CODATA), the Scientific Committee on Solar-Terrestrial Relations (SCOSTEP)and the Scientific Committee on Problems of the Environment (SCOPE). COSPAR has a Panel on Potentially Environmentally Detrimental Activities in Space (PEDAS) which has a great deal of common ground with IAU Commission 50 (Protection of Observatory Sites) in the mitigation of adverse environmental impacts on astronomy. However, in more recent times scientific partners have seen the need to look beyond the immense expertise of the ICSU family. One very successful collaboration has been between IAU Commission 50, the International Dark Skies Association and the *Commission International d'Eclairage* (CIE), in pursuing reduction of light pollution. This has resulted in CIE producing a zoning scheme to protect optical observatory sites – in particular those sites close to centres of population – and has created awareness among lighting engineers of the value of reducing light-spill, stimulating the design of new lighting schemes which use minimal energy and minimize light-spill. Some challenges, however, are national and international and necessarily involve interaction at the level of government – the topic of the next section.

5. The IAU and Governmental Organizations

The IAU has always sought to confine itself to the science of astronomy. ICSU does not permit items of a political nature to feature in the scientific meetings of its Unions, ICSU itself being the political interface. Politics destroyed ICSU's predecessor, the International Research Council. Eschewing political activity allows scientists from *all* nations to meet and to discuss their science. The IAU has successfully achieved this aim – not without several notable battles along the way – and has kept politics out of its meetings and activities. It is the great feature of IAU General Assemblies to see daily evidence of wide ranges of nationalities, not normally noted for being on warm speaking terms, vigorously disputing on astronomical matters. Absence of politics was an excellent policy when the IAU was founded; it is similarly excellent today. But there have been changes taking place in the last few decades which make the policy of steering clear of governmental organisations less absolute. Contact with intergovernmental institutions will demand information from the IAU in new and challenging formats. The necessary information may not exist in the optimum format for presentation nor, indeed, may it be readily available but will have to be sought. The IAU will have to finance the obtaining, analysis and preperation for presentation of those data.

The help of UNESCO in funding the Exchange of Astronomers and ISYA has already been mentioned in Sect. 3. Apart from UNESCO, however, there were few, if any, interactions with governmental organizations. In the last decade there has been a sea-change in outlook. ICSU has sought finance from UNEP (UN Environmental Program) for example for its environmental programmes.

More particularly for the IAU, the degradation of astronomical observing conditions by human activities has forced the search for partners beyond the ICSU family. The collaboration between IAU Commission 50 and CIE has already been mentioned. Again, UNESCO initially assisted the campaign against adverse environmental impacts on astronomy by funding an Exposition in 1992 at UNESCO in Paris to advertize, beyond the astronomical community, the severe problems experienced at all electromagnetic wavelengths by observational astronomy. This Exposition led to the publication of *The Vanishing Universe* (McNally 1994). The IAU was invited by the Scientific and Technical Sub-Committee of COPUOS to make presentations on adverse impacts on astronomy of activities in Space. In 1995, the IAU accepted the offer from COPUOS of Observer status on that Committee. The value of this status cannot be too heavily stressed since it offers opportunity for informed input to the Committee without compromising the IAU's independence. The IAU was invited by COPUOS to participate in UNISPACE III held in 1999 in Vienna by organising a symposium (IAU Symposium 196) on "Preserving the Night Sky" (Cohen and Sullivan 2001). The IAU therefore had an excellent opportunity to present to a wide audience – space scientists and engineers, educators, media and politicians *inter alia* – the entire spectrum of adverse environmental impacts on astronomy at a major international event. Importantly, the UN report on UNISPACE III reflected astronomical concerns and urged their amelioration. (Another outcome has already been mentioned in Sect. 4.) The connection with the UN COPUOS has proved immensely valuable and is likely to remain so in the foreseeable future. Closer contact might well be developed with the World Meteorological Organization. Significant climate change could affect some of the best observing sites and it is not too soon to study possible changes at trade-wind mountain sites.

IAU engagement with a wider range of ICSU bodies, non-ICSU, non-governmental and intergovernmental organisations may well increase in the future. As long as this engagement remains driven by astronomical imperatives, such wider scientific involvement can only be beneficial to astronomical science. However, the IAU will have new audiences to address and, as a consequence, have to meet new demands for astronomical information.

6. Information for the Public

The IAU is not primarily concerned with public relations; its founders envisaged, rather, an organization for international cooperation in some aspects of astronomical research. Early volumes of the *Transactions* show little evidence of concern for informing the public, that being left for individual astronomers to address in their repsective countries, as they saw fit. The growth of the astronomical profession since the World War II and the consequent growth of the IAU have changed things. Besides, astronomers in every country have become more dependent on public funds and public interest in astronomical discovery has greatly increased. Thus, the Union has had to engage to some extent in public relations. On the North-American continent, at least, the Central Bureau of Astronomical Telegrams has often acted, *de facto*, as the IAU's public relations office – as the Bureau's reports in the *Transactions* attest – mainly, of course, by providing information about transient bright objects, such as comets and novae.

In 1983 the *IB* had a cover added, on the back of which appeared a brief statement of the IAU's aims and responsibilities, clearly designed to inform the public rather than IAU members. Even then, the IAU was concerned to tell members of the public that star names "bought" from various commercial concerns would not be used by astronomers. That warning is, unfortunately, still needed; most astronomers have met at least one person who has either bought, or was thinking of buying, such a name and who was disappointed to learn that it would not be used. The IAU's efforts to raise public consciousness about light pollution and other environmental threats to astronomy have been discussed in Sect. 4 and 5. Both this matter and that of star names are discussed on the IAU web page (*www.iau.org*).

For IAU members, the web page has so far done little more than to change the way in which the Union disseminates information that it was already providing. In public relations, however, the web page has provided new opportunities. The section entitled "Frequently Asked Questions" is clearly aimed at the public. At the time of writing, in addition to the two topics already discussed, items will be found on: the status of Pluto, Near-Earth Objects, Planets around other Stars and the Spelling of Astronomical Names. Obviously, these items can be updated, added to or deleted as the public attention turns to other astronomical matters. We are perhaps witnessing the early stages of development of a new role for the IAU that has been made possible by the developments of information technology.

7. Conclusion

In this paper we have endeavoured to show the range of information services offered by the IAU. While specific examples have been highlighted, a paper such as this cannot be comprehensive. We have not even mentioned many Commission publications. We have listed the major formal publications of the IAU, looked at its collaborative role in providing information services in astronomy, its collaborations with other bodies to widen particular information bases and their availability beyond the astronomical community to other scientists, governments and the public. We have given an *in extenso* treatment of the IAU Telegrams because of their intimate connection with the IAU since its foundation to the present day. We hope that we have given a flavour of the manifold ways in which the IAU and its members have provided practical support to the science of astronomy since 1919 and are still planning services for the future. The IAU is moving into uncharted waters in its contacts with international governmental institutions, which may prove to be necessary and important, but we hope that the IAU will not lose sight of its primary historic obligation, to serve and support international astronomical science.

References

1. Arcetri Obs. (1926-39) *Immagini Spettroscopiche del Bordo Solare*, Appendices to *Osservazioni e Memorie del R. Oss. di Arcetri* **40-53**.
2. Batten, A.H. (1988) *Resolute and Undertaking Characters: The Lives of Wilhelm and Otto Struve*, Reidel, Dordrecht, Ch. 13.
3. Blaauw, A. (1994) *History of the IAU: The Birth and First Half-Century of the International Astronomical Union*, Kluwer Academic Publishers Dordrecht, Ch. 2 and 3.
4. Buchar, E. (1962) in *Trans. IAU*, **XIB**, (ed. D.H. Sadler), Academic Press, London and New York, pp. 73-4.
5. Cayrel, R., Smith, F.G., Fisher, A.J. & de Boer, J.B. (1980) *Guidelines for minimizing Urban Sky Glow near Observatories*, IAU/CIE Publication No. 1.
6. Cohen, R.J. & Sullivan, W.T. III (eds.) (2001) *Preserving the Astronomical Sky*, (IAU Symp. No. 196), ASP Conf. Series, San Francisco.
7. Delporte, E. (1930a) *Atlas Céleste (Report of Commission 3)*, Cambridge Univ. Press
8. Delporte, E. (1930b) *Délimitation Scientifique des Constellations*, Cambridge Univ. Press
9. Delporte, E. (1950) in *Trans. IAU* **VII**, (ed. J.H. Oort), Cambridge Univ. Press, p. 217.
10. Ferraz-Mello, S. (1983) in *Trans. IAU* **XVIIIB**, (ed. R.M. West), Reidel, Dordecht, pp. 305-7.
11. Fierro, J. (2001) in *Astronomy for Developing Countries* (IAU Special Session, ed. A.H. Batten), ASP Conf. Series, San Francsico, p. 178.
12. Fowler, A. (ed.) (1922) *Trans. IAU* **I**, Imperial College Bookstall, London, p. 207.
13. Gingerich, O. (1966) in *Trans. IAU* **XIIC**, *The Astronomers' Handbook*, (ed. J.-C. Pecker), Academic Press, London and New York, pp. 32-8.
14. Herget, P. (1960) in *Trans. IAU* **X**, (ed. D.H. Sadler), Cambridge Univ. Press, pp. 302-3.
15. Hers, J. (1982) in *Trans. IAU* **XVIIIA**, (ed. P.A. Wayman), Reidel, Dordrecht, p. 19.
16. Kholopov, P.N., Samus, N.N., Frolov, M.S., Goranskij, V.P., Gorynya, N.A., Kireeva, N.N., Kukarkina, N.P., Kurochkin, N.E., Medvedeva, G.I., Perova, N.B. &

Shugarov, S.Yu. (1985-95) *General Catalogue of Variable Stars*, Nauka, Moscow.
17. Lecointe, G. (1922) *Annuaire de l'Observatoire Royal de Belgique*, pp. 297-310.
18. Marsden, B. (1990) in *Trans. IAU* **XXB**, (ed. D. McNally), Kluwer, Academic Publishers, Dordrecht, p. 119.
19. McNally, D. (1994) *The Vanishing Universe*, Cambridge Univ. Press
20. Minnaert, M.G.J. (1966) in *Trans. IAU* **XIIB**, (ed. J.-C. Pecker), Academic Press, London and New York, pp. 629-649 and 43.
21. Oort, J.H. (ed.) (1950) in *Trans. IAU* **VII**, Cambridge Univ. Press, p. 28.
22. Pecker, J.-C. (ed.) (1966) *Trans. IAU* **XIIC**, *(The Astronomers' Handbook)*, Academic Press, London and New York, pp. 74-95 and 120-2.
23. Stephenson, F.R., Gurshtein, A., Dick, S.J. & Orchiston, W. (2002) in *IAU Information Bulletin* No. 90, pp. 18-19.
24. Thernöe, K.A. (1966) in *Trans. IAU* **XIIB** (ed. J.-C. Pecker), Academic Press, London and New York, pp. 78-9.
25. West, R.M. & Marsden, B.G. (1997) in *Trans. IAU* **XXIIIA**, (ed. I. Appenzeller), Kluwer Acad. Publ. Dordrecht, pp. 587-8.
26. Whipple, F.L. (1971) in *Trans. IAU* **XIVB**, (eds. C. de Jager & A. Jappel), Reidel, Dordrecht, p. 90.

WAS THE CARTE DU CIEL AN OBSTRUCTION TO THE DEVELOPMENT OF ASTROPHYSICS IN EUROPE?

D.H.P. JONES
Institute of Astronomy
Madingley Road
Cambridge CB3 0HA, UK
dhpj@ast.cam.ac.uk

Abstract. The Carte du Ciel project has been blamed for retarding the development of astrophysics in Europe by devouring scarce resources. Other factors are examined and found to be equally valid.

1. Introduction

It has sometimes been suggested (see *e.g.* Urban & Corbin 1998) that the Carte du Ciel project retarded the development of astrophysics in Europe leaving the field to be dominated by American observatories for more than half a century. It is inevitable that some fields of astronomy will develop faster in some countries than others and it is undeniable that a wide gap arose in extra-galactic astronomy between Europe and America in the first half of the 20th century. Véron (2001) has reviewed the progress of astrophysics in France and shown that the Carte du Ciel did indeed have a deleterious effect. However, the question posed bears the implicit assumption that astrophysics is more important than positional astronomy, overlooking their interdependence.

Scientifically this is a meaningless question because of the impossibility of performing a controlled experiment. Neither is there the possibility of making falsifiable predictions. Nevertheless it is interesting to review the progress of astrophysics during the period when the Carte du Ciel was active and review some alternative explanations.

A. Heck (ed.), Information Handling in Astronomy – Historical Vistas, 267-273.

2. History of Carte du Ciel Project

The Carte du Ciel was initiated at a conference in Paris in 1887 (Jones 2000). Prompted by the success of dry plate photography and the invention by the Henry brothers of an astrographic telescope with a well-corrected two-degree field, the conference decided on two parallel projects: first, to produce paper charts of the sky down to magnitude 14 and, second, to produce a precision astrographic catalogue of all stars brighter than 11. To complete the project in a reasonable time the work would be spread between eighteen observatories, each taking a zone of declination. Three astronomers of the 56 present were American but no American observatory took part in the project.

The Greenwich Astrographic telescope (Christie 1904) is shown in Fig. 1 and a list of participating observatories (Urban *et al.* 2001) in Table 1.

TABLE 1. The Declination Zone and Time Span for Each Observatory

Zone	Country	declination	limits	epoch
Greenwich	England	+90	+65	1892 1905
Vatican	Italy	+64	+55	1895 1922
Catania	Sicily	+54	+47	1894 1932
Helsinki	Finland	+46	+40	1892 1910
Potsdam	Germany	+39	+32	1893 1900
Hyderabad North	India	+39	+36	1928 1938
Uccle	Belgium	+35	+34	1939 1950
Oxford II	England	+33	+32	1930 1936
Oxford I	England	+31	+25	1892 1910
Paris	France	+24	+18	1891 1927
Bordeaux	France	+17	+11	1893 1925
Toulouse	France	+11	+5	1893 1936
Algiers	Algeria	+4	-2	1891 1912
San Fernando	Spain	-3	-9	1891 1918
Tacubaya	Mexico	-10	-16	1900 1938
Hyderabad South	India	-17	-23	1914 1929
Cordoba	Argentina	-24	-31	1909 1913
Perth	Australia	-32	-37	1902 1919
Perth-Edinburgh	Australia	-38	-40	1903 1914
Cape	South Africa	-41	-51	1897 1912
Sydney	Australia	-52	-64	1892 1948
Melbourne	Australia	-65	-90	1892 1928

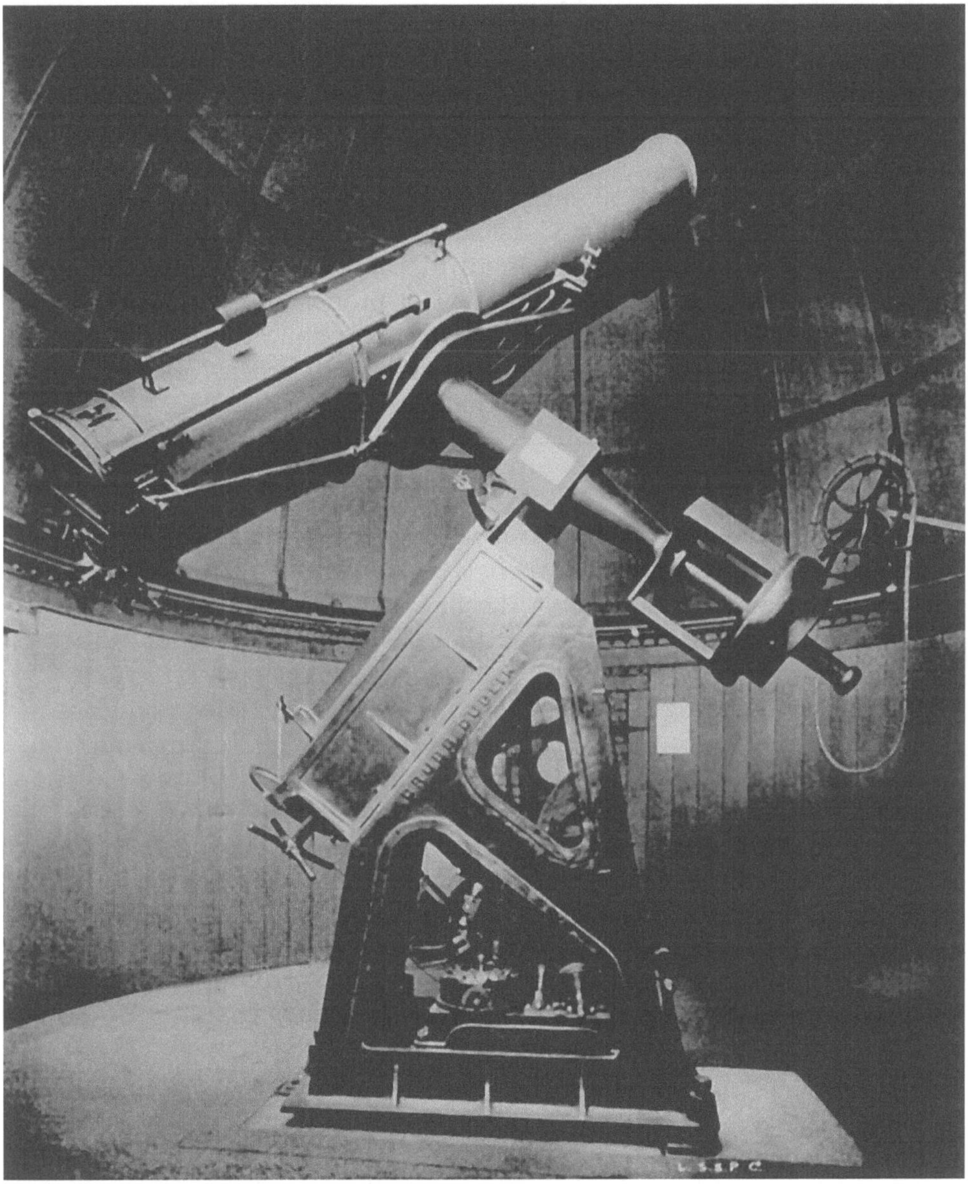

Figure 1. The Astrographic Telescope built in 1890 by Sir Howard Grubb for the Royal Observatory at Greenwich. All other Grubb telescopes followed the same design. Telescopes built by the brothers Henry in Paris employed English equatorial mounts and rectangular tubes.

The whole project was suspended for a year in 1900 to determine the solar parallax from the close opposition of Eros. Further delays were caused by World War I. Eventually the project was subsumed into the International Astronomical Union as Commission 23. The first report (Turner 1922) recorded that only two observatories, Greenwich and Oxford, had completed the observation, measurement and publication of their zones of the Astrographic Catalogue.

The project continued at much the same pace, interrupted by World War II, until the Astrographic Catalogue was completed and Commission 23 absorbed into Commission 24 in 1970 (Dieckvoss 1970). By this stage the main activity was the determination of proper motions, especially of variable stars. The 'Carte du Ciel' proper, the publication of charts, was never completed. In the early stages they were very expensive to print and latterly they were overtaken by the Palomar Sky Survey.

3. Large Telescopes

The history of the telescope has been well covered (King 1955). At the time of the Astrographic Conference there were already two large refractors in the United States: the 36-inch at the Lick and the 40-inch at the Yerkes Observatory. It was widely accepted that the practical limit for refractors had been reached.

For an increased light grasp it was necessary to use silver-on-glass reflectors, invented by von Steinheil and Foucault in 1856. In England, Common produced a 36-inch glass mirror which he sold to Edward Crossley. Crossley was never happy with this telescope and gave it to the Lick Observatory in 1895. Keeler used it with great success and a more robust mounting was built in 1904. Common also produced a 60-inch mirror which went first to Harvard College and then to the Armagh-Dunsink-Harvard observatory in Bloemfontein South Africa.

In 1862 it was proposed that a large reflector should be established at Melbourne in the Colony of Victoria (Warner 1982). The design was put in the hands of a committee in England who decided on a Cassegrain reflector with a 48-inch speculum. The silver-on-glass design was passed over as being unproven. As was usual, there were two specula, which alternated in the telescope while the other was being re-polished. This was done with a steam-driven polisher but it proved impossible to achieve an acceptable figure. Although a dome was designed, it was never built and wind buffeting was inevitable. The telescope made few significant observations and was generally felt to be a failure. Ritchey held that this fiasco delayed the development of large reflectors by thirty years *i.e.* until the demonstrable success of the Crossley reflector.

The Crossley reflector was followed by the Mount Wilson 60-inch (Adams 1954) in 1909 and the 100-inch in 1917. When the 200-inch arrived in 1948, glass had been replaced by Pyrex and silver by aluminium. By the end of the Carte du Ciel project all large telescope mirrors were of zero expansion ceramic.

4. Good Sites

Traditionally observatories were sited on some convenient hill near a town. Following a suggestion by Newton, Piazzi Smyth journeyed in 1856 to the heights of Tenerife and found that observing conditions were greatly improved at 10,700 ft. Lick Observatory on Mount Hamilton was the first deliberately sited on a previously tested mountain-top. Mount Wilson was chosen from a list of four sites for its excellent seeing and easy communications. Lick and several other American observatories set up observing stations in South America or South Africa on well-chosen sites.

In contrast most observatories in Europe were sited by historical accident and plagued by heavy cloud cover. Where European observatories had the opportunity to site telescopes in Africa they chose much better sites. In 1820 the British Admiralty founded an observatory at the Cape of Good Hope, intended to determine time to correct the chronometers of visiting ships. Happily it enjoys a mediterranean climate and it became a fruitful photometric site. In 1948, the Radcliffe Observatory opened its 74-inch Pyrex reflector in South Africa on a well-tested site in the High Veldt (Glass 1989). Seeing there is generally good and the least cloud cover is in winter, giving it a large number of clear observing hours. It quickly came to the forefront of observational cosmology. At the IAU General Assembly in 1952 (Hoyle 1954), Baade announced his doubling of the extra-galactic distance scale and was immediately followed by Thackeray with Radcliffe observations in good agreement. The 200-inch had been built to double the maximum distance observable with the 100-inch but it was now apparent that the 100-inch could see twice as far all along.

5. Benefactors

America was fortunate to attract many generous benefactors to astronomy *e.g.* Lick, Yerkes, Carnegie (the richest man in the world!) and Hooker. The mirror of the Mount Wilson 60-inch was a gift to Hale from his father.

European astronomy had its benefactors too. The Greenwich observatory received a twin refractor-reflector from the surgeon Sir H. Thompson, in 1897. In 1937 Greenwich received a 36-inch silver-on-glass Cassegrain reflector from the industrialist Mr. W.J. Yapp. Both telescopes were erected at Greenwich within a few yards of Flamsteed's 1675 observatory. Similarly the double refractor presented to the Cape by Sir Frank McClean was erected on the original site. The move of the Radcliffe Observatory was made possible by the philanthropist Sir William Morris. He wished to expand the Radcliffe Infirmary in Oxford and offered a handsome sum to buy the adjacent site occupied by the Radcliffe Observatory. The Radcliffe Observatory was an independent institution from the Oxford University Observatory where the Carte du Ciel had been carried out.

In the 19th century, the rich banker Raphaël Bischoffsheim presented the Paris Observatory with a meridian circle and an equatorial (Nath 2000). He went on to provide a 76-cm telescope with a 20-m dome for the Nice Observatory. It was sited on Mont Gros, chosen for the quality of the sky. Assan Dina was another benefactor of French astronomy who promoted site-testing campaigns in southern France in 1924 and 1925. They led to the establishment of l'Observatoire de Haute-Provence and its 1.93-m telescope, brought in to service in 1958. It was followed in 1960 by the 2-m Tautenberg Universal Reflecting Telescope in Germany, and in 1967 by the 2.5-m Isaac Newton Telescope in England. All these telescopes lay within their respective national boundaries. Europeans eventually gained access to world-class telescopes with the opening of ESO, at about the same time that the Carte du Ciel project was completed.

The Pic du Midi, at an altitude of 2890 m, in France enjoys excellent seeing but is very difficult to reach. The 2-m telescope was not commissioned until 1980, much later than the period discussed here.

6. Conclusions

Even if European astronomers had wished to pursue observational cosmology in the years 1920–1960 there would not have been enough people with the necessary interests, skills and training. Senior positions were held by people interested in theoretical, solar or positional astronomy and they tended to take students and hire junior staff with the same interests and attitudes. People who wished to pursue extra-galactic research, *e.g.* Baade, went to America.

It is concluded that the backwardness of extra-galactic astrophysics in Europe had several causes, among which the Carte du Ciel was only a minor one.

References

1. Adams, W.S. 1954, The Founding of the Mount Wilson Observatory *Publications of the Astronomical Society of the Pacific*, **66**, 267-303.
2. Christie, W.H.M. 1904, *Astrographic Catalogue 1900.0, Greenwich Section*, **Vol. I**, Plate I.
3. Dieckvoss, W. 1970, Report of Commission 23, *Trans. I.A.U.*, **XIVa**, 225-226.
4. Glass, I.S. 1989, The Story of the Radcliffe Telescope, *Q. Jl R. Astr. Soc.*, **30**, 33-58.
5. Hoyle, F. 1954, Report of Commission 28, *Trans. I.A.U.*, **VIII** , 397-398.
6. Jones, D.H.P. 2000, The Scientific Value of the Carte du Ciel *Astronomy & Geophysics*, **41**, 5.16-5.20.
7. King, H.C. 1955, *The History of the Telescope*, Sky Publishing Corporation, Cambridge, Massachusetts.
8. Nath, Al 2000, *Orion*, **58/5**, 38-39.
9. Turner, H.H. 1922, Report of Commission 23, *Trans. I.A.U.*, **I** , 60-66.
10. Urban, S.E. & Corbin, T.E. 1998, The Astrographic Catalogue, A Century of Work Pays Off, *Sky & Telescope*, **June 1998**, 41-44.
11. Urban, S.E., Corbin, T.E., Wycoff, G.L., Høg, E., Fabricius, C. & Makarov, V.V. 2001, *The AC 2000.2 Catalogue (CD-ROM)*, US Naval Obs., Washington & Copenhagen Univ. Obs., Copenhagen.
12. Véron, P. 2001, Préhistoire de l'Observatoire de Haute Provence, in *Observatoires et Patrimoine Astronomique Français*, Nantes, 8-9 juin 2001.
13. Warner, B. 1982, The large Southern Telescope: Cape or Melbourne?, *Q. Jl R. Astr.z Soc.*, **23**, 505-514.

AMATEUR DATA AND ASTRONOMICAL DISCOVERIES IN THE 20TH CENTURY

S. DUNLOP
140 Stocks Lane
East Wittering
Chichester PO20 8NT, UK
sdunlop@pact.cpes.susx.ac.uk

Abstract. Over the course of the last century, amateur astronomers have accumulated vast numbers of observational data, often of the highest scientific accuracy, on a wide range of astronomical objects. Amateur techniques have greatly advanced over that period and this paper summarizes some of the changes in various observational fields, and notes some of the significant scientific advances resulting from amateur work.

1. Introduction

Amateur data has always been important in astronomy, but the nineteenth century saw a distinct transition from astronomical research being largely carried out by private individuals (with both financial means and free time) to it becoming mainly the preserve of publicly-funded institutions and professional astronomers. Yet at the same time, there was a movement towards the establishment at both local and national level of groups of amateurs with a common interest in astronomy. The Société Astronomique de France[1] (SAF) was founded in 1887, the Astronomical Society of the Pacific[2] (ASP) in 1889, and the British Astronomical Association[3] (BAA) and the Royal Astronomical Society of Canada[4] (RASC) in 1890. Many societies, such as the ASP, RASC and SAF have always had substantial numbers of pro-

[1] http://www.iap.fr/saf/
[2] http://www.astrosociety.org/
[3] http://www.britastro.org/
[4] http://www.rasc.ca/

A. Heck (ed.), Information Handling in Astronomy – Historical Vistas, 275-294

fessional members. In Britain, because the Royal Astronomical Society[5] (RAS) had been formed much earlier (in 1820), the BAA's membership has always been primarily amateur, with a much smaller proportion of professionals than the RAS, where amateurs currently form about 55 per cent of the membership (Heck 2000).

Although many of the men and women who joined these societies were primarily concerned with the more recreational aspects (meeting and corresponding with like-minded people, attending lectures, and receiving publications devoted to astronomy) a significant number wanted to carry out serious observations and take part in collaborative work. For Great Britain and Ireland, the whole of this transitional period has been exceptionally well discussed and documented in a recent book (Chapman 1998).

Although there always has been, and presumably always will be, a place in astronomical research for dedicated individuals working alone, the foundation of national and international societies in the late nineteenth and early twentieth centuries was of fundamental importance in ensuring that amateurs were able to make a serious contribution to astronomical science. Their significance lay in the fact that, from the outset, certain societies set up, as part of their organisation, sections dedicated to particular fields of observation. Even more importantly, these observing sections not only encouraged individuals and provided them with general guidance in methods of observation, but also developed specific techniques appropriate to the fields concerned, and laid down explicit methods of recording and reporting observations. Some of these methods of visual observation were so well conceived that they have remained essentially unaltered to the present day, and are still producing scientifically valid results.

The majority of such groups also coordinated observations among observers and initiated specific observational programmes. Although collaboration had always existed, it had been carried out between friends or within relatively small groups of individuals. The advent of much larger groups, with members over a range of latitudes and longitudes, carrying out observations in a consistent manner, produced a much greater coverage, and a significant increase in the likelihood of detecting transient, previously unrecognized, phenomena.

In general, the individual observations in a specific field were submitted to a central coordinator (usually known as a Director, Secretary, or some similar title). From these observations a report was produced summarizing and analysing the activity, and this was published, normally annually, in one or other of the society's publications. The observations were then retained for form part of a central archive.

[5] `http://www.ras.org.uk/`

Some societies, most notably (in the early years) the British Astronomical Association, went far beyond just collating and summarizing the observational reports. Although the individual observers might undertake some reduction of their results, they were also required to submit the actual observations – as written in their observing notebooks or on specialised forms – so that direct comparisons with other observations could be made, and error checking could take place.

A specific example may illustrate this point. From the outset, the BAA's Variable Star Section asked observers to submit, not just the date and time of an observation, but the actual estimate – *i.e.*, the identification of the comparison stars used, and the actual numerical values estimated by the observer – as well as ancillary information about instrumentation, sky conditions, and the observer's assessment of the reliability of the actual estimate (not of the deduced magnitude). Observers would also submit a deduced magnitude as reduced from their actual estimate. Every single magnitude estimate was then checked to ensure that the mathematical reduction had used the correct magnitudes for the comparison stars, and had been carried out accurately. Taking the other information into account enabled an overall confidence level to be assigned to each and every deduced magnitude.

Such a rigorous procedure imposed an immense task upon the person or persons responsible for carrying out the checking (especially in the days before computers), but it did provide results of the highest scientific standard. Not only that, but the preservation of all the data in the archive meant that at any time the estimates could, if necessary, be re-reduced, if a comparison star was subsequently found to be variable or to have been assigned an incorrect magnitude. Similarly, the data could be re-examined by newly developed, or newly applied, statistical techniques.

Similar principles of scientific rigour were applied to other observational fields to a greater or lesser degree by many societies. As a result, long, consistent series of amateur data are available, covering, in some cases, more than one hundred years. In general, steps have been taken to ensure that any new observational methods that have been introduced over the years – typically photographic, photoelectric, or CCD techniques – are either consistent with the older visual methods, or form separate, consistent datasets.

Discussion of amateur work and amateur-professional collaboration may be found in various conference proceedings, in particular Dunlop & Gerbaldi (1988), Edberg (1992), and Percy & Wilson (2000). The high quality of much of the amateur observational work over the years has led to some notable scientific discoveries. The wide range of these and how they have been derived from the data are best illustrated by considering a selection of examples from different fields.

2. Aurorae – Upper atmosphere – NLC

Observation of aurorae has always been an amateur activity, although early in the century coordinated observations were largely confined to the Auroral Section of the BAA. Subsequently, the Finnish society, the Ursa Astronomical Association[6] (URSA) formed an active section in the 1920s. Other groups in Scandinavia, Russia, Japan, and North America started to contribute observations, particularly from mid-century onwards.

Amateur observations are largely confined to visual and photographic work (with some radio and magnetometer determinations of auroral activity). The data are analysed to determine the frequency and duration of displays, the forms present, heights, and other factors, including the relationship to solar activity. Amateur observations during the International Geophysical Year formed a significant portion of the data used by Akasofu in determining the extent and form of the auroral ovals and the episodic nature of auroral activity (Akasofu 1964). The significance of amateur observations has declined slightly with the advent of satellites that continuously monitor auroral activity, but still provides useful 'ground truth' and continuity with historical records.

The significance of amateur observations of noctilucent clouds (NLC) has, if anything, increased throughout the century with the apparent increase in the frequency of NLC for reasons that are currently unknown. Here again, visual and photographic records are typically analysed for frequency of occurrence, secular change, cyclic behaviour – a ten-year cycle is present – latitude, height, seasonal and daily variations. Without amateur observations, this field would be largely unknown (Gadsden 1998).

3. Meteors

Meteor studies were initially carried out by the sections within large societies such as the BAA and SAF, but the American Meteor Society[7] (AMS), founded in 1911, was one of the first organizations to be set up specifically to cover a single field of observation. Much later, the International Meteor Organization[8] (IMO) was formed in 1988 around a pre-existing nucleus of active European observers in Belgium, the Netherlands, and Germany.

In the early years much effort was expended in trying to attribute most meteors to specific radiants, often on the flimsiest of evidence. Early cata-

[6] `http://www.ursa.fi/`
[7] `http://www.amsmeteors.org/`
[8] `http://www.imo.net/`

logues included many suggested radiants, which are now known to be spurious. Typical of these are the many proposed by W.F. Denning, the BAA Meteor Section Director, just one of whose later lists may be cited as an example (Denning 1924).

Although many radiants have not stood the test of time, the actual observations, and the results deduced from them of shower or sporadic membership, rates, magnitudes, and distribution across streams now form an extensive archive for the analysis of long-term changes in meteor activity. Visual and photographic observations form the most significant portions of the cumulative archive, with lesser contributions from other methods of observation, including video, radio and, most recently, CCD material.

In the case of the Leonid meteor showers, amateur observations of recent peaks have largely confirmed as correct the predictions of rates and magnitudes made prior to the return, for example those of McNaught & Asher (1999).

In the early days of radio astronomy, the collaboration between Bernard Lovell and Manning Prentice, a solicitor and Director of the BAA Meteor Section, led to the correlation of radar echoes with meteors (Lovell 1968), culminating in the Giacobinid storm of 1946 (Lovell *et al.* 1947). Prentice was later to receive an honorary MSc for this work.

4. Fireballs

In general, observations of fireballs have been handled by the organisations dealing with meteors, although the data have naturally been analysed in a different way in an attempt to derive possible meteoroid orbits or meteorite fall sites. The majority of reports in such cases come from members of the public, so amateurs have played a lesser role. They have, however, participated in certain of the fireball networks set up from mid-century onwards, such as the Prairie Network, in the US Midwest, and the Canadian network, which achieved success, respectively, with the Lost City meteorite of 1970, and the Innisfree meteorite of 1977 (Halliday *et al.* 1978). Amateurs also play a prominent part in the current European Fireball Network[9], started in 1964 and still operating, coordinated by Ondrejov Observatory in the Czech Republic. One notable success for this network has been confirmation of the fireball of 2002 April 6 at 20:20:18 UT as a twin to the Pbram fall of 1959 April 7.

[9]`http://www.molau.de/meteore/imc97-2.html`

5. Meteorites

Although Jerzy Pokrzywnicki, a Polish lawyer, became an expert in the classification of meteorites (Turaj 1988) and published many papers on the subject, amateurs' main contribution have been in the recovery of specific meteorites or falls. One excellent example is the recent discovery by Joy of various strewn fields in Oman and the retrieval of nearly 200 fragments (Joy 2002).

6. Artificial satellites

Even before the first satellites were launched, Fred L. Whipple of the Smithsonian Astrophysical Observatory set up the Moonwatch organisation in 1955. Lasting for almost two decades, this vast band of about 5000 amateurs contributed about 400,000 observations of objects in orbit (Heyman 1996). Among radio observations, the group at Kettering Grammar School, led by Geoffrey Perry first tracked Sputnik 4 in May 1960. They subsequently went on to make a substantial contribution to knowledge of satellite launch sites, orbits, and decay times (Owen 1965), and the group expanded to include members in many different countries.

A large number of individuals in many different countries made visual observations of satellites. Notable among these was Russell Eberst of Edinburgh, the world's leading observer, who made over 90,000 observations in the period 1958-1982 (King-Hele 1983). Although many of the official programmes have been discontinued, there remains much work that can be done as detailed by Malley (2002). One group devoted to this work are the Visual Satellite Observers[10] (VSO).

7. Sun

Amateurs have always carried out intensive observations of the Sun, and most of the major societies carry reports of solar activity in their various publications. A few, such as the BAA, have tended to report on the active area count, rather than individual sunspot numbers. Observations from amateurs were included among those used by the Federal Observatory, Zurich to calculate Wolf (or Zurich) Numbers, until this programme ceased at the end of 1980. The Sunspot Index Data Centre[11] (SIDC) was founded in 1981 at the Royal Observatory of Belgium[12] to continue this work. The index is now known as the International Sunspot Number. An amateur's analysis showing an 80-year cycle has been presented by Gleissberg (1966).

[10] `http://www.satellite.eu.org/satintro.html`
[11] `http://sidc.oma.be/index.php3`
[12] `http://www.astro.oma.be/`

Several amateur observers have achieved impressive records. The late Herb Luft, an American observer, contributed observations over several decades, and the British observer Harold Hill has specialized in the observation of polar faculae. Many of the professional observatories obtain regular whole-disk photographs of the Sun, but because there is no real-time examination of the surface, interesting features are not examined in detail, and here the amateur observer scores. In addition, although many observatories obtain images through H-alpha (or similar) filters, if motion of a prominence (for example) shifts the light out of the passband, the feature will seem to have disappeared. Amateurs, using spectrohelioscopes, are able to compensate for the Doppler shift and thus continue to observe the evolution of the features, at the same time determining the velocity along the line of sight.

A description of a typical solar observing programme (that of the solar group of the German Vereinigung der Sternfreunde[13], VdS) may be found in the contribution by Reinsch (1988). This group has also published an excellent book on amateur astronomy of the Sun (Beck *et al.* 1995). An individual programme of observations is described by Hicks (2000).

8. Moon

Prior to the 1960s, a large amount of amateur work was carried out (particularly in Europe) on lunar cartography, with some notable Moon maps being produced, including the first with accurate positioning derived from photographs (Goodacre 1931), and a extremely detailed, large-scale map by Wilkins (Wilkins & Moore 1955), although some of the smaller or less prominent features mapped by Wilkins are not particularly trustworthy.

Considerable effort was also devoted to the study of transient phenomena, which became of particular importance prior to the Apollo landings (Middlehurst *et al.* 1968; Moore 1971), although the nature of these events is a subject of debate to the present day. With the mapping of most of the Moon by the Lunar Orbiter probes, amateur attention turned to the small, unmapped region near the South Pole, known as 'Luna Incognita' (Westfall 1991).

Despite the availability of detailed photographic mapping of the surface, a few dedicated observers have continued visual studies, producing highly detailed information about individual features, as shown, for example, by the work by Hill (1991). In general, however, the emphasis of amateur work has now passed to the study of limb features and the limb profile, particularly through occultation timings.

[13] `http://www.vds-astro.de/`

9. Occultations and transits

Although there had been considerable interest in the observation of lunar occultations over the years, with organisations such as the American Association of Variable Star Observers[14] (AAVSO) having an occultation section, it was not until the middle of the 20th century that this field of observation became popular, partly aided by the increasing availability of accurate time signals that could be received on portable equipment. Interest soon became concentrated on grazing occultations which could provide information about the profile of the lunar limb and which also occasionally revealed that an occulted star was a close, previously unrecognized, double. A considerable number of doubles have been found in this way (Provenmire 2002). The first prediction of an occultation of a minor planet was produced at this period (Taylor 1952b), together with the first prediction of an occultation by a minor planet (Taylor 1952a).

Initially this work was coordinated by the Nautical Almanac Office at the Royal Greenwich Observatory, Herstmonceux, but when this programme was forced to close, reduction was undertaken by the International Lunar Occultation Center[15] (ILOC) in Japan. Much of this work is now coordinated by the International Occultation Timing Association[16] (IOTA). An excellent summary of past and present occultation work is to be found in Dunham (2000). Currently occultations by the Moon and minor planets are studied by visual, photoelectric, and CCD techniques.

The study of occultation by minor planets has been a major success for amateurs, whose mobility meant that they were often able to secure useful observations by rushing to suitable sites when the orbit was refined shortly before the event. The shapes and sizes of many minor planets have been determined, one highly successful early result (but not the first) being that for Pallas (Taylor 1990; Dunham *et al.* 1990). It was amateur observations that also produced evidence for satellites of minor planets, although this was generally discounted before images from the Galileo spaceprobe revealed Dactyl as a satellite to (243) Ida. Occultation results also hinted at rings around Neptune, but these were only confirmed by observations from Voyager 2 in 1987.

In recent years, the great increase in sensitivity of equipment available to amateurs has meant that it has become possible to monitor some of the known extrasolar planetary systems for planetary transits. The first success in this field was recorded by the Astronomical Association Jyväskyän

[14] `http://www.aavso.org/`
[15] `http://www1.kaiho.mlit.go.jp/KOHO/iloc/docs/iloc_e.html`
[16] `http://www.lunar-occultations.com/iota/iotandx.htm`

Figure 1. The development of a dust storm on Mars in 2002 June to August. June 11: Meade 16-inch SCT at f/44; July 13 & August 19: 16-inch Newtonian at f/36. All images with Lynxx PC CCD camera, Koheisha R, G, B filters. (Don Parker, Coral Gables, Florida)

Sirius[17] in Finland, who detected the transit of the planet of HD 209458 (Oksanen 2002). A new organisation, Transitsearch[18] has just been set up, with the aim of coordinating amateur and professional work on such transits.

10. Planets

Serious amateur work on the planets may be said to have begun with the outstanding work by A.S. Williams on the motion of atmospheric features on Jupiter (Williams 1896). The monitoring of activity on Mars, Jupiter and Saturn continued throughout the twentieth century, primarily by visual means, with motion of features in longitude (and to a lesser extent, latitude) being tracked through the examination of drawings and central-meridian transit timings from many different observers.

Although he primarily used professional instruments at Meudon, E.M. Antoniadi was probably the greatest amateur Mars observer of the century. His observations resolved the nature of Martian surface features (Antoniadi 1909), as well as finally laying to rest the question of Martian 'canals' (Antoniadi 1915). His discovery of yellow clouds and the subsequent work by many other observers on Martian dust storms are fully covered by McKim (1999). Recent detection of 'flares' from Edom Promontorium on Mars – presumed to be sunlight glinting on ice crystals – show that even well-studied planets still repay observation (Chandler 2002).

[17] `http://nyrola.jklsirius.fi/`
[18] `http://www.transitsearch.org/`

Understanding of the meteorology of Jupiter was primarily advanced by members of the Association of Lunar and Planetary Observers[19] (ALPO), BAA, and SAF and (from roughly mid-century onwards), the Società Astronomica Italiana[20] (SAI) and Société Astronomique de Suisse[21] (SAS). Some notable contributions were P.B. Molesworth's discovery of the South Tropical Disturbance in 1901 (McKim 1997); T.E.R. Phillips' observations that led to Peek's recognition of the circulation (the Circulating Current) at that latitude (Peek 1926; Phillips 1939); the finding that there were one or more discrete sources of activity below the visible cloud deck (Reese 1953); the work by Rogers and Wacker on global upheavals in the 1970s and 1980s, and evidence of cyclic activity in the North Equatorial Belt. Early amateur results are summarized in Peek's classic work (Peek 1958), and comprehensive discussion of subsequent amateur and professional observations is given by Rogers (1995).

Saturn offers fewer opportunities for amateur discoveries than Jupiter but J.C. Bartlett and W.H. Haas discovered the bicoloured aspect of the rings (Alexander 1949), and analysis of the work of various observers revealed the existence of seasonal changes on the planet (McKim & Blaxall 1984).

In general, amateur photography of the planets has been a useful adjunct to visual observations, with some remarkable work by amateurs such as Jean Dragesco and Don Parker which has revealed as much detail as visual observations and greatly aided interpretation of results. With the introduction of CCD equipment, however, some truly outstanding images have been obtained, so that amateur surveillance is likely to play an increasingly important role in future (Figs. 1 & 2). One major success for conventional photography was the discovery by Charles Boyer, a magistrate, working from Brazzaville, of the approximately 4-day retrograde rotation of the atmosphere of Venus, first reported in 1960 (Boyer & Camichel 1960).

11. Planetary satellites

By their very nature, satellites of the planets offer few opportunities for amateur study, but one outstanding visual result was the determination of Ganymede's captured rotation (Phillips 1922).

Timings of mutual phenomena (occultations or eclipses) of satellites may be used for astrometric purposes to refine both satellite sizes and orbital parameters. The principal observational campaigns have been or-

[19] `http://www.lpl.arizona.edu/alpo/`
[20] `http://www.sait.it/`
[21] `http://www.astroinfo.ch/sag/index_sag_f.html`

Figure 2. Jupiter, 2002 Oct. 30, 04:53 UT, 16-inch Newtonian at f/21.4, Lynxx PC CCD camera, Koheisha R, G, B filters. (Don Parker, Coral Gables, Florida)

ganized by the Bureau des Longitudes[22] in Paris, including PHEMU85 (Arlot 1988), PHEMU91 (Arlot *et al.* 1997), PHEMU97 and PHEMU03 for Jupiter; PHESAT95 (Arlot *et al.* 1996) for Saturn; and the forthcoming 2007-2008 campaign for satellites of Uranus.

Stellar occultations by planetary satellites offer another interesting field for amateurs. The results that may be obtained are admirably illustrated by the extensive discussion of the occultation of 28 Sgr by Titan on 1989 July 3 (Hollis & Miles 1994). This revealed small-scale structure in Titan's atmosphere, evidence for dispersion of the central flash as a function of wavelength, and even set an upper limit on the diameter of 28 Sgr. The occultation of HIP 106829 by Titania on 2001 September 8, produced a refined position for Uranus itself, an upper limit on the atmosphere of Titania, and the angular diameter of the occulted star (Sicardy 2002).

12. Minor planets

For the first half of the 20th century, amateur work on minor planets was largely limited to the occasional accidental discovery. Gradually, however, photography came to be used for astrometry and the refinement of orbits, and this work was generally carried out though specialized sections (sometimes one devoted to computation) within the major organisations. Some societies, such as ALPO and the BAA, have specific Minor Planet (Asteroid) sections.

[22] `http://www.imcce.fr/`

A significant increase in amateur involvement came with the first predictions of occultations by minor planets (as described earlier), with the initial successes soon proving their worth. With the growth in photoelectric methods in the 1980s, increasing numbers of rotational light-curves (and thus periods) were determined. The real impetus came with the introduction of CCD astrometry, however, with amateurs now discovering large numbers of new minor planets, some as a result of specific survey programmes and others as a by-product of supernova searches. With suitable equipment, amateurs are now able to reach such low magnitudes that even research into Kuiper-Belt objects is feasible (Offutt 2002) – and is carried out.

With better photometry of surrounding fields, even visual estimates of magnitudes (generally used for the production of phase-curves) still make a very useful contribution.

13. Comets

Amateurs have always played a considerable part in the observation of comets, with a large number of reports of activity being given in the various societies' publications over the years. In the cases of the brighter comets, some reports have been extremely extensive, with detailed drawings of structures such as jets and hoods within the coma, as well as tail structures and disconnection events (Bergé *et al.* 1988; Caron *et al.* 1988). Apart from European and North-American observers, Japanese amateurs have been very active in this field, and a useful short summary of their work is given by Hasegawa (1988).

Over the years, amateurs have discovered a number of comets, some serendipitously, and some as a result of deliberate search programmes. Generally such searches have been undertaken individually, rather than coordinated by a cometary group, and this was especially the case with the surge in discoveries in Japan, following the introduction of prizes for new discoveries. The most prolific observer, however, was William Bradfield, in Australia, who discovered 17 comets in the years between 1972 and 1995. Currently, although most comets are found by professional searches, amateurs continue to make significant discoveries every year.

A peak of amateur cometary activity came with the return of Comet P/Halley and the formation of the International Halley Watch (IHW). This programme (which included work on Comet P/Crommelin and Comet P/Giacobini-Zinner) saw 1575 registered participants, and resulted in an enormous archive of material (Sekanina & Fry 1991; Edberg 1996). An analytical summary of the implications of such programmes is given by Edberg (2000). A current programme is the International Ulysses Comet Watch

Network[23], which is monitoring comets in conjunction with the European Space Agency's Ulysses spaceprobe, with a particular emphasis on disconnection events.

14. Variable stars

The study of variable stars is one field where amateur observations have been (and still remain) supremely successful. Strangely, this has not always been accepted by some professionals. Strohmeier (1972), for example, speaks scathingly of amateur visual work, despite (perhaps unknowingly) using data derived from visual work by amateurs himself. Most professionals, however, recognize that amateur observations have been vital in many individual fields, such as the study of long-period and semiregular variables and, most particularly, cataclysmic variables (Warner 1995).

There are a number of societies specifically devoted to the study of variable stars, most notable of which is the American Association of Variable Star Observers (AAVSO), which was founded in 1911. Some of the other prominent societies or groups are: Association Française d'Observateurs d'Étoiles Variables[24] (AFOEV); Bundesdeutsche Arbeitsgemeinschaft für Veränderliche Sterne[25] (BAV); Variable Star Section of the British Astronomical Association[26] (BAA-VSS); Variable Star Section of the Royal Astronomical Society of New Zealand[27] (RASNZ); Astronomical Society of Southern Africa[28] (ASSA).

These societies' variable-star archives hold vast numbers of magnitude estimates. The AAVSO archive alone amounts to over 10 million estimates; that for the RASNZ Variable Star Section, about 4 million; and the BAA-VSS, nearly 2 million. When the observations of amateurs, obtained before the various societies were founded, are taken into account, many stars have runs of data extending for well over a century. There are observations of U Geminorum, for example, from its discovery on 1855 December 5 to the present day.

Discoveries from amateur observations of variables are too numerous to mention. Various contributions and descriptions of amateur-professional collaborations (such as the amateur input to the Hipparcos mission) may be found in Percy (1986), Dunlop & Gerbaldi (1988), Edberg (1992), Percy *et al.* (1992), and Percy & Wilson (2000). One particularly outstanding discovery was made by Bailey, who subsequently became a professional

[23] `http://lasp.colorado.edu/ucw/aboutucw.html`
[24] `http://astro.u-strasbg.fr/afoev`
[25] `http://www.bav-astro.de/`
[26] `http://www.britastro.org/vss/`
[27] `http://www.rasnz.org.nz/`
[28] `http://www.assa.za.org/frames.html`

astronomer, when he was able to demonstrate a relationship between the decay timescale of dwarf novae and their orbital period (Bailey 1975).

Although observations were originally made purely by visual means, and this method is still widely used, photographic, photoelectric, and CCD methods have also been employed to greater or lesser extents. The complexities of using photoelectric methods in this and other fields prompted the formation in 1980 of the International Amateur-Professional Photoelectric Photometry[29] (IAPPP) group. Although photoelectric methods are giving way to CCD photometry, several individual groups of amateurs have successfully developed and regularly use automated telescopes to make full use of the equipment's potential. CCD techniques are now making a very major contribution as regards discoveries, photometry of variable-star fields, and actual magnitude determinations.

Numerous objects are kept under surveillance by amateurs in connection with either permanent or short-term observational programmes in collaboration with professional astronomers. Alerts are normally channelled through the major organisations, such as the AAVSO and *The Astronomer* [30] (TA). The best groups ensure that any reports are checked and confirmed before issuing specific alerts.

Large numbers of new, or suspected, variable stars, novae and supernovae have been discovered by amateurs in the course of their work, and such reports are normally channelled through the individual organisations. In the first half of the 20th century, novae were largely discovered by accident, the epitome being perhaps the discovery of Nova Puppis 1942 (CP Pup) by the Japanese schoolgirl Kuniko Sofue at 03:00 local time, while she was darning socks (Shigehisa 2000). From the middle of the century onwards, amateurs began to carry out coordinated visual and photographic patrols for novae and supernovae and, more recently, have had numerous successes with CCD equipment. This has been particularly noticeable in the field of supernova discoveries, where the numbers being found by amateurs rival those found by professional search programmes. At the time of writing, Tom Boles and Mark Armstrong of the UK Nova/Supernova Patrol (for example) have found 25 and 39 supernovae, respectively, on CCD images.

The search programmes bring other results. Also in connection with the UK Nova/Supernova Patrol, Mike Collins, using simple photographic means over an approximately 12-year period, has discovered or recovered 175 variables[31], with 32 being currently listed in the *General Catalogue of Variable Stars* (GCVS), and 55 suspected variables being included in

[29] `http://www.iappp.vanderbilt.edu/`

[30] `http://www.theastronomer.org/`

[31] `http://www.theastronomer.org/mikes_variables.html`

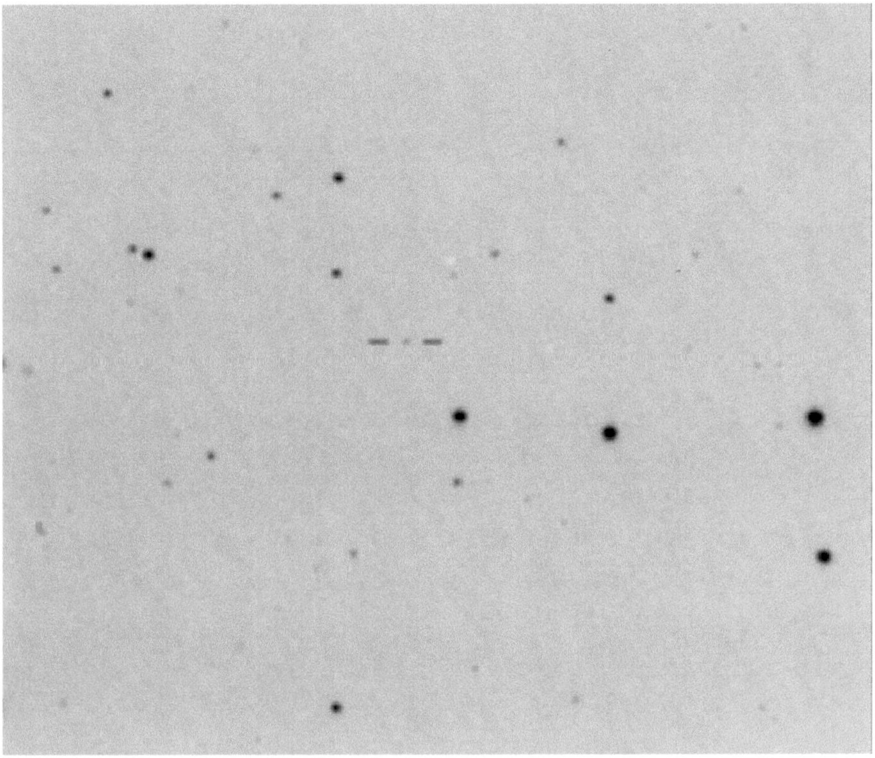

Figure 3. GRB 021004 on 2002 Oct.04, 21:23 UT, composite of 10 30-second images. Celestron 14 SCT at F/11, SBIG ST-9e CCD camera. (Mark Armstrong, Rolvenden, Kent, UK)

the *New Catalogue of Suspected Variable Stars* (NSV). Approximately 150 other stars are pending confirmation. Amateur-professional collaborative programmes such as the All Sky Automated Survey (ASSA) are producing phenomenal numbers of observations and new discoveries (di Cicco 2002).

With the observation by members of the Astronomical Association Jyväskyän Sirius[32] in Finland of the afterglow from a gamma-ray burst (Kato 2000; Oksanen 2002), and several amateurs' recent success in 2002 with GBB 021004, (Figs. 3 & 4) it has become obvious that, despite the problems, close amateur-professional collaboration, such as within the GRB Coordinates Network[33] (GCN), can bring more success in this challenging field.

[32]`http://nyrola.jklsirius.fi/`
[33]`http://gcn.gsfc.nasa.gov/`

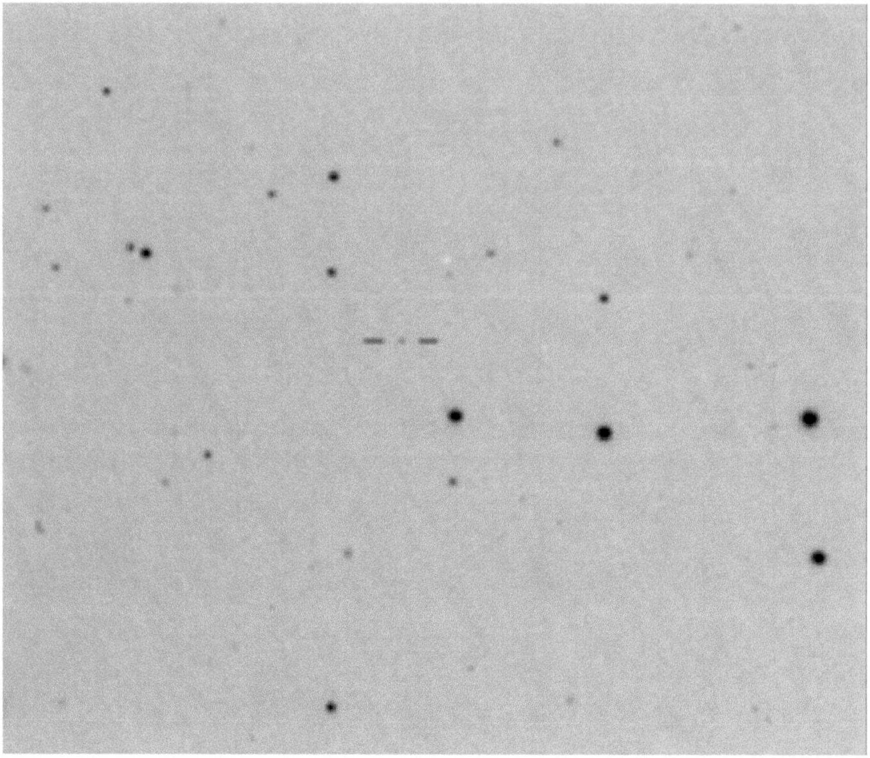

Figure 4. GRB 021004 on 2002 Oct.05, 01:10 UT, single 60-second image. Celestron 14 SCT at F/11, SBIG ST-9e CCD camera. (Mark Armstrong, Rolvenden, Kent, UK)

15. Double stars

Early in the 20th century, a few amateurs were prominent in the discovery of double stars, with S.W. Burnham, T.E. Espin, R. Jonckheere, M.V. Duruy, and P. Baise contributing many examples (Couteau 1981 & 1988). Although there was a general decline in interest in mid-century, the subject was kept alive, particularly by the efforts of members of the Double Star Section of the SAF, which had a direct input to the Inca-Hipparcos programme (Durand 1988). Experimentation with photographic and other methods began in the 1980s (Rodriguez *et al.* 1988), and a number of new doubles have been found from lunar occultations (Provenmire 2002). There has recently been a revival of interest in doubles (Tanguay 1999) and although the introduction of CCD instrumentation has not brought such benefits as in other fields, notably because of limitations on resolution and limiting magnitude, the application of such techniques suggests that a full revival may be under way (Anton 2002).

16. Conclusion

This survey makes no pretence to be complete, and amateurs have been active in a number of other areas. One interesting case is that of the amateur photographer, M. de Kerolyr, who, working with professional instruments, using long exposures, and combining several images, managed to record details of nebulosity not shown by any other workers, before or since (Richardot 1988a&b). A number of similar examples of unique contributions could be cited.

One point that may be emphasized is that, with the growth of practically instantaneous global communications, it is essential that observations and discoveries be properly verified before being made available to the astronomical community at large. Certain amateurs have always, regrettably, made hasty claims for 'discoveries', and unfortunately professional astronomers have diverted valuable telescope time on the basis of spurious accounts. Much time has been wasted by both the amateur and professional communities in checking and attempting to verify such reports. The activities of such organisations as *The Astronomer*, which rigorously verifies all reports before disseminating them widely, is one that should be followed by all amateur organisations.

In addition, it should also be borne in mind that material on the World Wide Web is, by its very nature, ephemeral, and that therefore there is no adequate substitute for the observations being submitted to a central archive, and published in full or as analyses in a peer-reviewed journal. At present, although the SIMBAD[34] database maintained by the Centre de Données astronomiques de Strasbourg (CDS) includes references to papers in peer-reviewed amateur publications, it neither includes data on, nor links to, amateur archives.

If these points are borne in mind, there can be no doubt that amateur data will long continue to provide both the raw material for research and new discoveries in astronomy for a long time in the future.

References

1. Akasofu, S.I. 1964, The development of the auroral substorm, *Planet. Space Sci.* **12**, 273.
2. Alexander, A.F.O'D. 1949, Saturn Section Council Report for 1948-49, *J. Brit. astron. Assoc.* **59**, 216-217.
3. Anton, R. 2002, Measuring Double Stars with Video, *Sky & Telescope* **104/July**, 117-120.

[34]`http://cdsweb.u-strasbg.fr/simbad/`

4. Antoniadi, E.M. 1909, Mars Section Report, *J. Brit. astron. Assoc.* **20**, 78-81.
5. Antoniadi, E.M. 1915, Mars in 1909, *Mem. Brit. astron. Assoc.* **20**, pt 2.
6. Arlot, J.E. 1988, Amateurs, Contribution to the PHEMU85 Campaign Observing Mutual Phenomena of Jupiter's Satellites, in *Stargazers: The Contribution of Amateurs to Astronomy*, Eds. S. Dunlop & M. Gerbaldi, Springer-Verlag, 198.
7. Arlot, J.E. *et al.* 1997, A catalogue of the observations of the mutual phenomena of the Galilean satellites made in 1991 during the PHEMU91 campaign, *Astron. Astrophys. Suppl.* **125**, 399-405.
8. Arlot, J.E. *et al.* 1996, First results of the PHESAT95 campaign of observation of the phenomena of the satellites of Saturn, in *Dynamics and astrometry of natural and artificial celestial bodies*, Proc. IAU Colloq. 165, Poznan.
9. Bailey, J. 1975, Rate of decline from dwarf nova outbursts, *J. Brit. astron. Assoc.* **86**, 30 & 74.
10. Beck, R. *et al.* 1995, Solar Astronomy Handbook, Willmann-Bell, Richmond Virginia (first published as Handbuch der Sonnenbeobachter, Vereinigung der Sternfreunde, Berlin, Bonn, 1982)
11. Bergé, P.M. *et al.* 1988, P/Halley: The Disconnection Event of 1986 April 11, in *Stargazers: The Contribution of Amateurs to Astronomy*, Eds. S. Dunlop & M. Gerbaldi, Springer-Verlag, 155-157.
12. Boyer, C. & Camichel, H. 1960, Étude photographique de Vénus en lumière violette et ultra-violette, *L'Astronomie* **74**, 375-381.
13. Caron, R. *et al.* 1988, A Study of Disconnection Velocities in the Plasma Tail of P/Halley, in *Stargazers: The Contribution of Amateurs to Astronomy*, Eds. S. Dunlop & M. Gerbaldi, Springer-Verlag, 158-159.
14. Chandler, D.L. 2002, Mars Odyssey finds signs of ice at flare site, *Sky & Telescope* **104/August**, 26.
15. Chapman, A. 1998, The Victorian Amateur Astronomer: Independent Astronomical Research in Britain, 1820-1920, Wiley-Praxis, Chichester.
16. Couteau, P. 1981, Observing Visual Double Stars, Trans. A. Batten, MIT Press, Cambridge MA, 10-11.
17. Couteau, P. 1988, The Historical Contribution of Amateurs to the Study of Double Stars, in *Stargazers: The Contribution of Amateurs to Astronomy*, Eds. S. Dunlop & M. Gerbaldi, Springer-Verlag, 15-19.
18. Denning, W.F. 1924, Radiant points of shooting stars observed at Bristol chiefly from 1912 to 1922 inclusive, *Mon. Not. Roy. Astron. Soc.* **84**, 43-56.
19. di Cicco, D. 2002, Big Variable-Star Payoff from Small Scopes, *Sky & Telescope* **104/October**, 18-19.
20. Dunham, D.W. *et al.* 1990, The Size and Shape of (2) Pallas from the 1983 Occultation of 1 Vulpeculae, *Astron. J.* **99 (5)**, 1636.
21. Dunham, D.W. *et al.* 2000, Amateur Professional Partnerships in the Observation of Occultations, in *Amateur-Professional Partnerships in Astronomy*, Eds. J.R. Percy & J.B. Wilson, *Astron. Soc. Pacific Conf. Series* **200**.
22. Dunlop, S. & Gerbaldi, M. (Eds.) 1988, Stargazers: The Contribution of Amateurs to Astronomy, Proc. IAU Colloq. 98, Springer-Verlag.
23. Durand, P. 1988, The SAF Double Star Section's Contribution to the INCA-HIPPARCOS Programme, in *Stargazers: The Contribution of Amateurs to Astronomy*, Eds. S. Dunlop & M. Gerbaldi, Springer-Verlag, 93-94.
24. Edberg, S.J. 1996, The Archive of the Amateur Observation Network of the International Halley Watch, Vol. 1: Comet Giacobini-Zinner + Vol. 2: Comet Halley, *Jet Propulsion Lab. Publ.* **96-3**, Vols. 1 & 2, Pasadena CA.
25. Edberg, S.J. (Ed.) 1992, Research Amateur Astronomy, *Astron. Soc. Pacific Conf. Series* **33**.
26. Edberg, S.J. 2000, Amateur-Professional Cooperation: Lessons Learned from the International Halley Watch 1981-1989, in *Amateur-Professional Partnerships in As-*

tronomy, Eds J.R. Percy & J.B. Wilson, *Astron. Soc. Pacific Conf. Series* **200**, 108-112.

27. Gadsden, M. 1998, The North-West Europe data on noctilucent clouds: a survey, *J. Atmos. & Solar-TerrPhys.* **60**, 1163-1174.
28. Gleissberg, W. 1966, Ascent and descent of the eighty-year cycles of solar activity, *J. Brit. astron. Assoc.* **76**, 265.
29. Goodacre, W. 1931, The Moon: with a Description of its Surface Form Fully Illustrated by the Author's Revised Map of the Moon in Twenty-five Sections, together with Numerous Drawings, Charts and Photographs, published by the author, Bournemouth.
30. Halliday, I., Blackwell, A.T. & Griffin, A.A. 1978, The Innisfree meteorite and the Canadian camera network, *J. Roy. Astron. Soc. Canada* **72**, 15-39.
31. Hasegawa, I. 1988, Observations of Comets, Minor Planets and Meteors in Japan, in *Stargazers: The Contribution of Amateurs to Astronomy*, Eds. S. Dunlop & M. Gerbaldi, Springer-Verlag, 87-89.
32. Heck, A. 2000, Characteristics of Astronomy-Related Organizations, *Astrophys. Sp. Sc.* **274**, 733-783.
33. Heyman, I.M. 1996, Smithsonian Perspectives, *Smithsonian Magazine* **April**.
34. Hicks, J. 2000, The Many Faces of the Sun: Observing & Photographing the Solar Chromosphere Through a Solar Cycle, in *Amateur-Professional Partnerships in Astronomy*, Eds. J.R. Percy & J.B. Wilson, *Astron. Soc. Pacific Conf. Series* **200**, 146-156.
35. Hill, H. 1991, A portfolio of lunar drawings, Cambridge University Press, Cambridge.
36. Joy, W. 2002, personal communication.
37. Kato, T. 2000, GRB 000926, optical observation by Oksanen, *GCN GRB Obs. Rep.* **813**, Kyoto University.
38. King-Hele, D. 1983, Observing Earth Satellites, Macmillan London, 107-108.
39. Lovell, A.C.B. 1968, The Story of Jodrell Bank, Oxford Univ. Press, 8-10.
40. Lovell, A.C.B. *et al.* 1947, Radio echo observations of the Giacobinid meteors 1946, *Mon. Not. Roy. Astron. Soc.* **107**, 164-175.
41. McKim, R.J. 1999, Telescopic Martian Dust Storms: A Narrative and Catalogue, *Mem. Brit. astron. Assoc.* **44**.
42. McKim, R.J. 1997, P.B.Molesworth's discovery of the great South Tropical Disturbance on Jupiter, 1901, *J. Brit. astron. Assoc.* **107**, 239-245.
43. McNaught, R.H. & Asher D.J. 1999, Leonid Dust Trails and Meteor Storms, *WGN* **27**, 85-102.
44. Malley, P.D. 2000, Partnership Proposal Between the Amateur and Professional Communities in the Field of Earth Satellite Observing, in *Amateur-Professional Partnerships in Astronomy*, Eds. J.R. Percy & J.B. Wilson, *Astron. Soc. Pacific Conf. Series* **200**, 142-145.
45. Middlehurst, B.A. *et al.* 1968, Chronological Catalog of Reported Lunar Events, *NASA Technical Report* **TR R-277**, Washington, DC.
46. Moore, P.A. 1971, Extension of the Chronological Catalogue of Reported Lunar Events, October 1967-June 1971, *J. Brit. astron. Assoc.* **81**, 365.
47. Offutt, W.B. 2000, Kuiper Belt Objects: Observing, From an Amateur's Perspective, in *Amateur-Professional Partnerships in Astronomy*, Eds. J.R. Percy & J.B. Wilson, *Astron. Soc. Pacific Conf. Series* **200**, 123-130.
48. Oksanen, A. 2002, Amateur Astrophysics in the Arctic Circle, *Sky & Telescope* **104/November**, 67-73.
49. Owen, K. 1965, Kettering's Cosmos Scholars, *Flight International* (Special issue, 22 July).
50. Peek, B.M. 1926, Abnormal Jovian Spots in 1920, *J. Brit. astron. Assoc.* **37**, 62-64.
51. Percy, J.R. (Ed.) 1986, The Study of variable stars using small telescopes, Cambridge Univ. Press.

52. Percy, J.R., Mattei, J.A. & Sterken, C. 1992, Variable Star Research: An international perspective, Cambridge Univ. Press, Cambridge.
53. Percy, J.R. & Wilson, J.B. 2000, Amateur-Professional Partnerships in Astronomy, *Astron. Soc. Pacific Conf. Series* **200**.
54. Peek, B.M. 1958, The Planet Jupiter, Faber & Faber, London.
55. Phillips, T.E.R. 1922, The Third Satellite of Jupiter, *Mem. Roy. astron. Soc.* **LXXXII**, 96-100.
56. Phillips, T.E.R. 1939, Saturn Section Report, *Mem. Brit. astron. Assoc.* **34**, pt 2.
57. Provenmire, H. 2002, The newly discovered binary star 44 Eta Librae and other binary stars discovered by lunar occultations, *IAPP Comm.* **81** (September 2000 [sic]).
58. Reese, E.J. 1953, A Possible Clue to the Rotation Period of the Solid Nucleus of Jupiter, *J. Brit. astron. Assoc.* **63**, 219-221.
59. Reinsch, K. 1988, The Sonne Network – Amateur Solar Astronomy in Germany, in *Research Amateur Astronomy*, *Astron. Soc. Pacific Conf. Series* **33**, 22-31.
60. Richardot, A. 1988a, A Renowned, Yet Forgotten Astronomer: Marcel de Kerolyr, in *Stargazers: The Contribution of Amateurs to Astronomy*, Eds. S. Dunlop & M. Gerbaldi, Springer-Verlag, 58-61.
61. Richardot, A. 1988b, Un amateur célèbre et oublié: Marcel de Kerolyr, Observations et Travaux, Société Astronomique de France, Hors-Série, No. 1 (Juin).
62. Rodriguez, D. *et al.* 1988, An Improved Method of Photographic Measures of Double Stars, in *Stargazers: The Contribution of Amateurs to Astronomy*, Eds. S. Dunlop & M. Gerbaldi, Springer-Verlag, 123-127.
63. Rogers, J.H. 1995, The Giant Planet Jupiter, Cambridge Univ. Press, Cambridge.
64. Sekanina, Z. & Fry, L. 1991, The Comet Halley Archive Summary, *Jet Propulsion Lab. Publ.* **400-450**, Pasadena CA.
65. Shigehisa, O. 2000, The Story of the "Socks Star" – A Bright Nova Discovered by a Schoolgirl in War Time, in *Amateur-Professional Partnerships in Astronomy*, Eds. J.R. Percy & J.B. Wilson, *Astron. Soc. Pacific Conf. Series* **200**.
66. Sicardy, B. 2002, What we learned about Titania, *Sky & Telescope* **104/July**, 94.
67. Strohmeier, W. 1972, Variable Stars, Pergamon Press, Oxford, 5.
68. Tanguay, R.C. 1999, Observing Double Stars for Fun and Science, *Sky & Telescope* **97/February**, 116-121.
69. Taylor, G.E. 1952b, Occultations of asteroids by the Moon, *J. Brit. astron. Assoc.* **62**, 177-178.
70. Taylor, G.E. 1952b, An Occultation by a Minor Planet, *J. Brit. astron. Assoc.* **62**, 176-177.
71. Taylor, G.E. 1990, The size and shape of (2) Pallas from the 1983 occultation of 1 Vulpeculae, *Astron. J.* **99**, 1636.
72. Turaj, K. 1988, Amateur Astronomy in Poland: Past and Present, in *Stargazers: The Contribution of Amateurs to Astronomy*, Eds. S. Dunlop & M. Gerbaldi, Springer-Verlag, 40-43.
73. Warner, B. 1995, Cataclysmic Variable Stars, Cambridge Univ. Press, 5-6.
74. Westfall, J.E. 1991, The Luna Incognita Project, *Sky & Telescope* **82**, 556-557, 559.
75. Wilkins, H.P. & Moore, P. 1955, The Moon: A Complete Description of the Surface of the Moon, Containing the 300-inch Wilkins Lunar Map, Faber & Faber, London.
76. Williams, A.S. 1896, On the Drift of the Surface Material of Jupiter in Different Latitudes, *Mon. Not. Roy. astron. Soc.* **LVI**, 143.

Previously published in Astrophysics and Space Science Library book series:

- **Volume 282**: **Radio Recombination Lines**
  Authors: M.A. Gordon, R.L. Sorochenko
  Hardbound, ISBN 1-4020-1016-8, November 2002
- **Volume 280**: **Organizations and Strategies in Astronomy III**
  Editor: André Heck
  Hardbound, ISBN 1-4020-0812-0, September 2002
- **Volume 279**: **Plasma Astrophysics , Second Edition**
  Author: Arnold O. Benz
  Hardbound, ISBN 1-4020-0695-0, July 2002
- **Volume 277**: **The Sun and Space Weather**
  Author: Arnold Hanslmeier
  Hardbound, ISBN 1-4020-0684-5, July 2002
- **Volume 273**: **Lunar Gravimetry**
  Author: Rune Floberghagen
  Hardbound, ISBN 1-4020-0544-X, May 2002
- **Volume 272**:**Merging Processes in Galaxy Clusters**
  Editors: L. Feretti, I.M. Gioia, G. Giovannini
  Hardbound, ISBN 1-4020-0531-8, May 2002
- **Volume 271**: **Astronomy-inspired Atomic and Molecular Physics**
  Author: A.R.P. Rau
  Hardbound, ISBN 1-4020-0467-2, March 2002
- **Volume 270**: **Dayside and Polar Cap Aurora**
  Authors: Per Even Sandholt, Herbert C. Carlson, Alv Egeland
  Hardbound, ISBN 1-4020-0447-8, July 2002
- **Volume 269**:**Mechanics of Turbulence of Multicomponent Gases**
  Authors: Mikhail Ya. Marov, Aleksander V. Kolesnichenko
  Hardbound, ISBN 1-4020-0103-7, December 2001
- **Volume 268**:**Multielement System Design in Astronomy and Radio Science**
  Authors: Lazarus E. Kopilovich, Leonid G. Sodin
  Hardbound, ISBN 1-4020-0069-3, November 2001
- **Volume 267: The Nature of Unidentified Galactic High-Energy Gamma-Ray Sources**
  Editors: Alberto Carramiñana, Olaf Reimer, David J. Thompson
  Hardbound, ISBN 1-4020-0010-3, October 2001
- **Volume 266: Organizations and Strategies in Astronomy II**
  Editor: André Heck
  Hardbound, ISBN 0-7923-7172-0, October 2001
- **Volume 265: Post-AGB Objects as a Phase of Stellar Evolution**
  Editors: R. Szczerba, S.K. Górny
  Hardbound, ISBN 0-7923-7145-3, July 2001
- **Volume 264: The Influence of Binaries on Stellar Population Studies**
  Editor: Dany Vanbeveren
  Hardbound, ISBN 0-7923-7104-6, July 2001
- **Volume 262: Whistler Phenomena**
  Short Impulse Propagation
  Authors: Csaba Ferencz, Orsolya E. Ferencz, Dániel Hamar, János Lichtenberger
  Hardbound, ISBN 0-7923-6995-5, June 2001

- **Volume 261: Collisional Processes in the Solar System**
  Editors: Mikhail Ya. Marov, Hans Rickman
  Hardbound, ISBN 0-7923-6946-7, May 2001
- **Volume 260: Solar Cosmic Rays**
  Author: Leonty I. Miroshnichenko
  Hardbound, ISBN 0-7923-6928-9, May 2001
- **Volume 259: The Dynamic Sun**
  Editors: Arnold Hanslmeier, Mauro Messerotti, Astrid Veronig
  Hardbound, ISBN 0-7923-6915-7, May 2001
- **Volume 258: Electrohydrodynamics in Dusty and Dirty Plasmas Gravito-Electrodynamics and EHD**
  Author: Hiroshi Kikuchi
  Hardbound, ISBN 0-7923-6822-3, June 2001
- **Volume 257: Stellar Pulsation - Nonlinear Studies**
  Editors: Mine Takeuti, Dimitar D. Sasselov
  Hardbound, ISBN 0-7923-6818-5, March 2001
- **Volume 256: Organizations and Strategies in Astronomy**
  Editor: André Heck
  Hardbound, ISBN 0-7923-6671-9, November 2000
- **Volume 255: The Evolution of the Milky Way Stars versus Clusters**
  Editors: Francesca Matteucci, Franco Giovannelli
  Hardbound, ISBN 0-7923-6679-4, January 2001
- **Volume 254: Stellar Astrophysics**
  Editors: K.S. Cheng, Hoi Fung Chau, Kwing Lam Chan, Kam Ching Leung
  Hardbound, ISBN 0-7923-6659-X, November 2000
- **Volume 253: The Chemical Evolution of the Galaxy**
  Author: Francesca Matteucci
  Hardbound, ISBN 0-7923-6552-6, May 2001
- **Volume 252: Optical Detectors for Astronomy II State-of-the-art at the Turn of the Millennium**
  Editors: Paola Amico, James W. Beletic
  Hardbound, ISBN 0-7923-6536-4, December 2000
- **Volume 251: Cosmic Plasma Physics**
  Author: Boris V. Somov
  Hardbound, ISBN 0-7923-6512-7, September 2000
- **Volume 250: Information Handling in Astronomy**
  Editor: André Heck
  Hardbound, ISBN 0-7923-6494-5, October 2000
- **Volume 249: The Neutral Upper Atmosphere**
  Author: S.N. Ghosh
  Hardbound, ISBN 0-7923-6434-1, July 2002
- **Volume 247: Large Scale Structure Formation**
  Editors: Reza Mansouri, Robert Brandenberger
  Hardbound, ISBN 0-7923-6411-2, August 2000
- **Volume 246: The Legacy of J.C. Kapteyn Studies on Kapteyn and the Development of Modern Astronomy**
  Editors: Piet C. van der Kruit, Klaas van Berkel
  Hardbound, ISBN 0-7923-6393-0, August 2000

- **Volume 245: Waves in Dusty Space Plasmas**
  Author: Frank Verheest
  Hardbound, ISBN 0-7923-6232-2, April 2000
- **Volume 244: The Universe**
  Visions and Perspectives
  Editors: Naresh Dadhich, Ajit Kembhavi
  Hardbound, ISBN 0-7923-6210-1, August 2000
- **Volume 243: Solar Polarization**
  Editors: K.N. Nagendra, Jan Olof Stenflo
  Hardbound, ISBN 0-7923-5814-7, July 1999
- **Volume 242: Cosmic Perspectives in Space Physics**
  Author: Sukumar Biswas
  Hardbound, ISBN 0-7923-5813-9, June 2000
- **Volume 241: Millimeter-Wave Astronomy: Molecular Chemistry & Physics in Space**
  Editors: W.F. Wall, Alberto Carramiñana, Luis Carrasco, P.F. Goldsmith
  Hardbound, ISBN 0-7923-5581-4, May 1999
- **Volume 240: Numerical Astrophysics**
  Editors: Shoken M. Miyama, Kohji Tomisaka, Tomoyuki Hanawa
  Hardbound, ISBN 0-7923-5566-0, March 1999
- **Volume 239: Motions in the Solar Atmosphere**
  Editors: Arnold Hanslmeier, Mauro Messerotti
  Hardbound, ISBN 0-7923-5507-5, February 1999
- **Volume 238: Substorms-4**
  Editors: S. Kokubun, Y. Kamide
  Hardbound, ISBN 0-7923-5465-6, March 1999
- **Volume 237: Post-Hipparcos Cosmic Candles**
  Editors: André Heck, Filippina Caputo
  Hardbound, ISBN 0-7923-5348-X, December 1998
- **Volume 236: Laboratory Astrophysics and Space Research**
  Editors: P. Ehrenfreund, C. Krafft, H. Kochan, V. Pirronello
  Hardbound, ISBN 0-7923-5338-2, December 1998
- **Volume 235: Astrophysical Plasmas and Fluids**
  Author: Vinod Krishan
  Hardbound, ISBN 0-7923-5312-9, January 1999
  Paperback, ISBN 0-7923-5490-7, January 1999
- **Volume 234: Observational Evidence for Black Holes in the Universe**
  Editor: Sandip K. Chakrabarti
  Hardbound, ISBN 0-7923-5298-X, November 1998
- **Volume 233: B[e] Stars**
  Editors: Anne Marie Hubert, Carlos Jaschek
  Hardbound, ISBN 0-7923-5208-4, September 1998
- **Volume 232: The Brightest Binaries**
  Authors: Dany Vanbeveren, W. van Rensbergen, C.W.H. de Loore
  Hardbound, ISBN 0-7923-5155-X, July 1998
- **Volume 231: The Evolving Universe**
  Selected Topics on Large-Scale Structure and on the Properties of Galaxies
  Editor: Donald Hamilton
  Hardbound, ISBN 0-7923-5074-X, July 1998

- **Volume 230: The Impact of Near-Infrared Sky Surveys on Galactic and Extragalactic Astronomy**
  Editor: N. Epchtein
  Hardbound, ISBN 0-7923-5025-1, June 1998
- **Volume 229: Observational Plasma Astrophysics: Five Years of Yohkoh and Beyond**
  Editors: Tetsuya Watanabe, Takeo Kosugi, Alphonse C. Sterling
  Hardbound, ISBN 0-7923-4985-7, March 1998
- **Volume 228: Optical Detectors for Astronomy**
  Editors: James W. Beletic, Paola Amico
  Hardbound, ISBN 0-7923-4925-3, April 1998
- **Volume 227: Solar System Ices**
  Editors: B. Schmitt, C. de Bergh, M. Festou
  Hardbound, ISBN 0-7923-4902-4, January 1998
- **Volume 226: Observational Cosmology with the New Radio Surveys**
  Editors: M.N. Bremer, N. Jackson, I. Pérez-Fournon
  Hardbound, ISBN 0-7923-4885-0, February 1998
- **Volume 225: SCORe'96: Solar Convection and Oscillations and their Relationship**
  Editors: F.P. Pijpers, Jørgen Christensen-Dalsgaard, C.S. Rosenthal
  Hardbound, ISBN 0-7923-4852-4, January 1998
- **Volume 224: Electronic Publishing for Physics and Astronomy**
  Editor: André Heck
  Hardbound, ISBN 0-7923-4820-6, September 1997
- **Volume 223: Visual Double Stars: Formation, Dynamics and Evolutionary Tracks**
  Editors: J.A. Docobo, A. Elipe, H. McAlister
  Hardbound, ISBN 0-7923-4793-5, November 1997
- **Volume 221: An Atlas of Local Group Galaxies**
  Authors: Paul W. Hodge, Brooke P. Skelton, Joy Ashizawa
  Hardbound, ISBN 1-4020-0673-X, November 2002
- **Volume 222: Remembering Edith Alice Müller**
  Editors: Immo Appenzeller, Yves Chmielewski, Jean-Claude Pecker, Ramiro de la Reza, Gustav Tammann, Patrick A. Wayman
  Hardbound, ISBN 0-7923-4789-7, February 1998
- **Volume 220: The Three Galileos: The Man, The Spacecraft, The Telescope**
  Editors: Cesare Barbieri, Jürgen H. Rahe†, Torrence V. Johnson, Anita M. Sohus
  Hardbound, ISBN 0-7923-4861-3, December 1997

Missing volume numbers have not yet been published.
For further information about this book series we refer you to the following web site: http://www.wkap.nl/prod/s/ASSL

To contact the Publishing Editor for new book proposals:
Dr. Harry (J.J.) Blom: harry.blom@wkap.nl

MIX
Papier aus verantwortungsvollen Quellen
Paper from responsible sources
FSC® C105338

If you have any concerns about our products,
you can contact us on
ProductSafety@springernature.com

In case Publisher is established outside the EU,
the EU authorized representative is:
Springer Nature Customer Service Center GmbH
Europaplatz 3, 69115 Heidelberg, Germany

Printed by Libri Plureos GmbH
in Hamburg, Germany